Terence N. Mitchell · Burkhard Costisella

NMR – From Spectra to Structures

Springer

Berlin
Heidelberg
New York
Hong Kong
London
Milan
Paris
Tokyo

Terence N. Mitchell · Burkhard Costisella

NMR – From Spectra to Structures

An Experimental Approach

Springer

Terence N. Mitchell
Universität Dortmund
– Fachbereich Chemie –
44221 Dortmund
Germany
e-mail: mitchell@chemie.uni-dortmund.de

Burkhard Costisella
Universität Dortmund
– Fachbereich Chemie –
44221 Dortmund
Germany
e-mail: Costi@chemie.uni-dortmund.de

ISBN 3-540-40695-6 Springer-Verlag Berlin Heidelberg New York

Cataloging-in-Publication Data applied for
Bibliographic information published by Die Deutsche Bibliothek
Die Deutsche Bibliothek lists this publication in the Deutsche Nationalbibliografie;
detailed bibliographic data is available in the Internet at <http://dnb.ddb.de>.

Springer-Verlag Berlin Heidelberg New York
a member of BertelsmannSpringer Science + Business Media GmbH

http://www.springer.de

Printed in Germany

Typesetting: medio Technologies AG, Berlin
Cover design: Künkel/Lopka, Heidelberg

Printed on acid-free paper 5 4 3 2 1 0

dedicated to Reiner Radeglia
an NMR pioneer in a then divided Germany

Preface

Why write another NMR book? Most of the many already available involve theoretical approaches of various kinds and levels of complexity. Few books deal with purely practical aspects and a handful are slanted towards problem-solving. Collections of problems of different complexity are invaluable for students, since theory of itself is not very useful in deducing the structure from the spectra.

However, there is now a huge variety of NMR experiments available which can be used in problem-solving, in addition to the standard experiments which are a "must". We start by providing an overview of the most useful techniques available, as far as possible using one single molecule to demonstrate which information they bring. The problems follow in the second part of the book.

Readers can obtain a list of answers to the problems on application (by mail or e-mail) to the authors.

We thank Annette Danzmann, Christa Nettelbeck and Bernhard Griewel for their invaluable help in recording the spectra and our wives Karin and Monika for their patience and support during the writing of the book. We also thank Bernd Schmidt for reading the manuscript and giving us valuable tips on how it could be improved. Finally we thank the staff at Springer for turning our manuscript into the finished product you now have in your hands.

Terence N. Mitchell
Universität Dortmund
– Fachbereich Chemie –
44221 Dortmund
Germany
e-mail:
mitchell@chemie.uni-dortmund.de

Burkhard Costisella
Universität Dortmund
– Fachbereich Chemie –
44221 Dortmund
Germany
e-mail:
Costi@chemie.uni-dortmund.de

Table of Contents

Introduction

NMR spectroscopy is arguably the most important analytical method available today. The reasons are manifold: it is applied by chemists and physicists to gases, liquids, liquid crystals and solids (including polymers). Biochemists use it routinely for determining the structures of peptides and proteins, and it is also widely used in medicine (where it is often called MRI, Magnetic Resonance Imaging). With the advent of spectrometers operating at very high magnetic fields (up to 21.1 T, i.e. 900 MHz proton resonance frequency) it has become an extremely sensitive technique, so that it is now standard practice to couple NMR with high pressure liquid chromatography (HPLC). The wide range of nuclei which are magnetically active makes NMR attractive not only to the organic chemist but also to the organometallic and inorganic chemist. The latter in particular often has the choice between working with liquid or solid samples; the combination of high resolution and magic angle spinning (HR/MAS) of solid samples provides a wealth of structural information which is complementary to that obtained by X-ray crystallography. The same suite of techniques, slightly adapted, is now available to those working in the field of combinatorial chemistry. This is only a selection of the possibilities afforded by NMR, and the list of methods and applications continues to multiply.

No single monograph can hope to deal with all the aspects of NMR. In writing this book we have concentrated on NMR as it is used by preparative chemists, who in their day-to-day work need to determine the structures of unknown organic compounds or to check whether the product obtained from a synthetic step is indeed the correct one.

Previous authors have taught the principles of solving organic structures from spectra by using a combination of methods: NMR, infrared spectroscopy (IR), ultraviolet spectroscopy (UV) and mass spectrometry (MS). However, the information available from UV and MS is limited in its predictive capability, and IR is useful mainly for determining the presence of functional groups, many of which are also visible in carbon-13 NMR spectra. Additional information such as elemental analysis values or molecular weights is also often presented.

It is however true to say that the structures of a wide variety of organic compounds can be solved using only NMR spectroscopy, which provides a huge arsenal of measurement techniques in one to three dimensions. To determine an organic structure using NMR data is however not always a simple task, depending on the complexity of the molecule. This book is intended to provide the necessary tools for solving organic structures. However, it does not just consist of a series of problems. These form Part 2 of the book, and in Part 1 a relatively simple organic compound (**1**) is used as an example to present the most important 1D and 2D experiments.

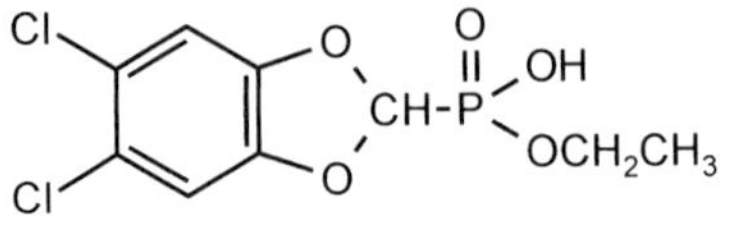

1

All the magnetic nuclei present in the molecule (^{1}H, ^{13}C, ^{31}P, ^{17}O, ^{35}Cl) are included in the NMR measurements, and the necessary theory is discussed very briefly: the reader is referred to suitable texts which he or she can consult in order to learn more about the theoretical aspects.

The molecule which we have chosen will accompany the reader through the different NMR experiments; the "ever-present" structure will make it easier to understand and interpret the spectra.

Our standard molecule is however not ideally suited for certain experiments (e.g. magnetic non-equivalence, NOE, HPLC-NMR coupling). In such cases other simple compounds of the same type, compounds **2-7**, will be used:

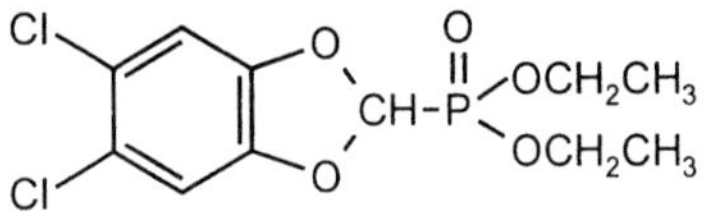

2 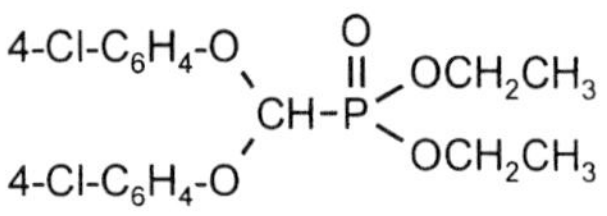 **3**

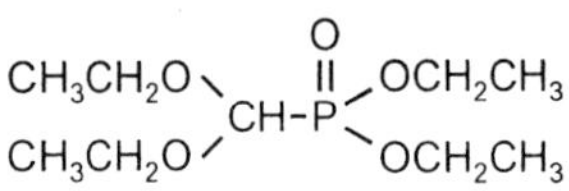

4 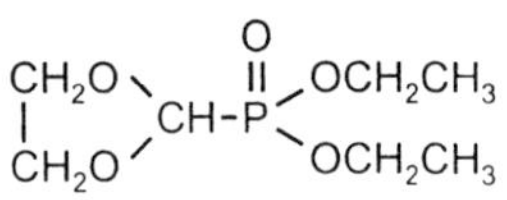 **5**

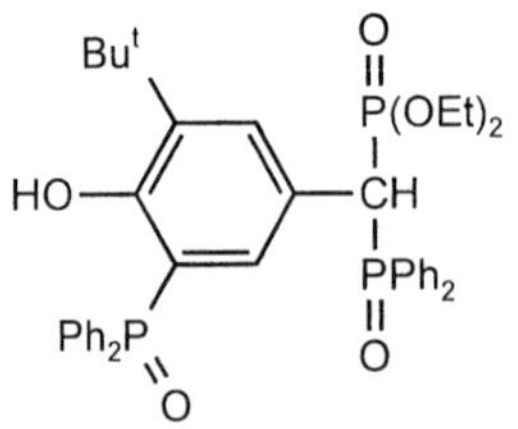

6 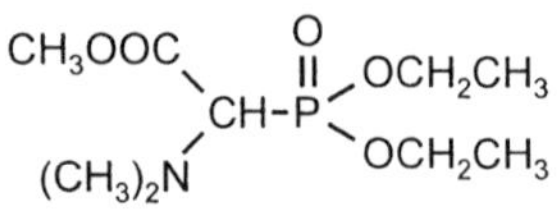 **7**

Part 1: NMR Experiments

This book is not intended to teach you NMR theory, but to give you a practical guide to the standard NMR experiments you will often need when you are doing structure determination or substance characterisation work, and (in Part 2) to provide you with a set of graded problems to solve. At the end of Part 1 we shall recommend some books which you will find useful when you are working on the problems.

Thus we shall try to take you through Part 1 without recourse to much theory. We shall of course use many terms which will be unfamiliar to you if you have not yet had a course in NMR theory, and these will be emphasised by using **bold** lettering when they appear. You can then, if you wish, go to the index of whatever theory book you have available in order to find out exactly where you can read up on this topic. From time to time, when we feel it advisable to say one or two words about more theoretical aspects in our text, we shall do so using *italic.*

The Appendix at the end of Part 1 contains a list of recommended texts for theoretical and experimental aspects of NMR as well as for solving spectroscopic problems.

1 1D Experiments

1.1 ^{1}H, D (^{2}H): Natural Abundance, Sensitivity

Hydrogen has two NMR-active nuclei: ^{1}H, always known as "the proton" (thus "proton NMR"), making up 99.98%, and ^{2}H, normally referred to as D for deuterium.

These absorb at completely different frequencies, and since deuterium and proton chemical shifts are identical (also because deuterium is a **spin-1 nucleus**), deuterium NMR spectra are hardly ever measured.

However, NMR spectrometers use deuterium signals from deuterium-labelled molecules to keep them stable; such substances are known as **lock substances** and are generally used in the form of solvents, the most common being deuterochloroform $CDCl_3$.

1.1.1
Proton NMR Spectrum of the Model Compound 1

Before we start with the actual experiment it is very important to go through the procedures for preparing the sample. The proton spectra are normally measured in 5 mm sample tubes, and the concentration of the solution should not be too high to avoid line broadening due to viscosity effects. For our model compound we dissolve 10 mg in 0.6 mL $CDCl_3$: between 0.6 and 0.7 mL solvent leads to optimum **homogeneity**. It is vital that the solution is free from undissolved sample or from other insoluble material (e.g. from column chromatography), since these cause a worsening of the homogeneity of the magnetic field. Undesired solids can be removed simply by filtration using a Pasteur pipette, the tip of which carries a small wad of paper tissue.

The sample is introduced into the spectrometer, locked onto the deuterated solvent (here $CDCl_3$) and the homogeneity optimised by **shimming** as described by the instrument manufacturer (this can often be done automatically, particularly when a sample changer is used).

The proton experiment is a so-called **single channel experiment**: the same channel is used for sample irradiation and observation of the signal, and the irradiation frequency is set (automatically) to the resonance frequency of the protons at the magnetic field strength used by the spectrometer.

Although some laboratories have (very expensive) spectrometers working at very high fields and frequencies, routine structure determination work is generally carried out using instruments whose magnetic fields are between 4.6975 Tesla (proton frequency 200 MHz) and 14.0296 Tesla (600 MHz). *The NMR spectroscopist always characterises a spectrometer according to its proton measuring frequency!*

The precise measurement frequency varies slightly with solvent, temperature, concentration, sample volume and solute or solvent polarity, so that exact adjustment must be carried out before each measurement. This process, known as **tuning and matching**, involves variation of the capacity of the circuit. Modern spectrometers carry out such processes under computer control.

The measurement procedure is known as the **pulse sequence**, and always starts with a delay prior to switching on the irradiation pulse. The irradiation pulse only lasts a few microseconds, and its length determines its power. The NMR-active nuclei (here protons) absorb energy from the pulse, generating a signal.

To be a little technical: the magnetisation of the sample is moved away from the y-axis, and it is important to know the length of the so-called 90° pulse which, as the name suggests, moves it by 90°, as this is needed in other experiments. In the experiment we are discussing now, a shorter pulse (corresponding to a pulse angle of 30-40°, the so-called ***Ernst angle****) is much better than a 90° pulse.*

When the pulse is switched off, the excited nuclei return slowly to their original undisturbed state, giving up the energy they had acquired by excitation. This process is known as **relaxation**. The detector is switched on in order to record the decreasing signal in the form of the **FID** (free induction decay). You can observe the FID on the spectrometer's computer monitor, but although it actually contains all the information about the NMR spectrum we wish to obtain, it appears completely unintelligible as it contains this information as a function of time, whereas we need it as a function of frequency.

This sequence, delay-excitation-signal recording, is repeated several times, and the FIDs are stored in the computer. The sum of all the FIDs is then subjected to a mathematical operation, the **Fourier transformation**, and the result is the conventional NMR spectrum, the axes of which are frequency (in fact chemical shift) and intensity. Chemical shift and intensity, together with coupling information, are the three sets of data we need to interpret the spectrum.

Figure 1 shows the proton spectrum of our model compound, recorded at a frequency of 400 MHz.

All signals are assigned to the corresponding protons in the molecular formula: this is made easier by prediction programmes. Table 1 presents the result of a prediction compared with the actual values.

If you do not have a prediction programme available, look on the Internet to see whether you can find freeware or shareware there. Otherwise use tables such as those you will find in the book by Pretsch et al. (see Appendix).

We shall now consider these signals and demonstrate the correctness of the assignment using different NMR techniques.

Before continuing, remind yourself of the rules for **spin-spin coupling**, i.e. for determining the number of lines in a multiplet (with the help of the "**n+1 rule**") and their intensities (using **binomial coefficients**).

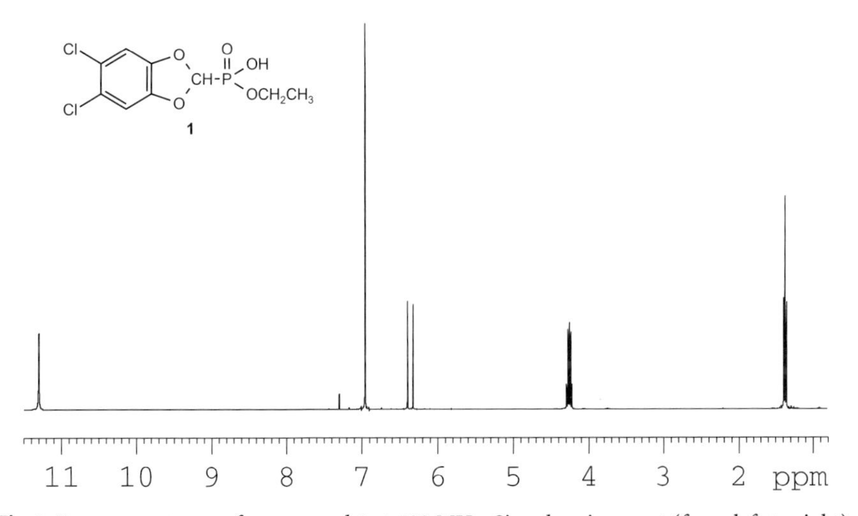

Fig. 1 Proton spectrum of compound 1 at 400 MHz. Signal assignment (from left to right): OH proton (singlet), aromatic protons (singlet), methine proton (doublet), OCH_2 protons (apparently a quintet), CH_3 protons, triplet. The small signal at 7.24 ppm is due to $CHCl_3$.

Table 1

Chemical shift (ppm)	J_{HP} (Hz)	Chemical shift (calc.)	J_{HP} (calc.)	Assignment
11.58	0	10.6	0	OH
6.92	not observed	7.0	0.3	CH_{arom}
6.32	28.7	6.6	16.9	CH-P
4.20	8.0	4.2	8.4	CH_2
1.33	0.6	1.3	1.0	CH_3

Let us start with the one-line signal on the left, the singlet, at 11.58 ppm. Our standard, tetramethylsilane TMS, gives a one-line signal whose chemical shift is defined as 0.00 ppm. Signals to its left are said to absorb at low field, those to its right (quite unusual in fact) at high field of TMS. Thus the signal at 11.58 ppm is that which absorbs at the lowest field, and we have assigned this as being due to the OH proton. This proton is acidic, the O-H bond being relatively weak, and can thus undergo fast chemical exchange with other water molecules or with deuterated water, D_2O. Thus if our sample is treated with 1-2 drops of D_2O and shaken for a few seconds the OH signal will disappear when the spectrum is recorded again: a new signal due to HOD appears at 4.7 ppm.

This technique works for any acidic proton present in a compound under investigation and is very useful in structure determination.

The next signal is a very small one at 7.24 ppm and comes from the small amount of $CHCl_3$ present in the $CDCl_3$.

The singlet at 6.92 ppm is due to the two aromatic protons: these have identical environments and thus show no coupling with other protons. They are too far from the phosphorus atom to show measurable coupling to it.

The two lines between 6.25 and 6.40 ppm are in fact a doublet due to the methine (CH) proton, which absorbs at relatively low field because it is bonded to two electronegative oxygen atoms. This proton is very close (separated by only two bonds) to the phosphorus, which is a **spin-1/2 nucleus** (there is only one isotope, phosphorus-31). The proton is also a spin-1/2 nucleus, so that H-H and H-P coupling behaviour is analogous. The distance between the two lines in the doublet is the coupling constant J, or to be exact $^2J_{P\text{-}C\text{-}H}$ and must be given in Hz, *not* ppm! The actual J value is 28.7 Hz.

How can we show that the two lines are due to a coupling? We need to carry out a so-called **decoupling** experiment, which "eliminates" couplings. Since two different nuclei are involved here, we do a **heterodecoupling** experiment (as opposed to **homodecoupling** when only one type of nucleus is involved, most commonly the proton). Decoupling is a 2-channel experiment in which we excite (and observe) the protons with channel 1 and excite the phosphorus nuclei with channel 2, which we call the decoupling channel. Channel 2 is set to the phosphorus resonance frequency, which we can obtain from tables; the excitation of the phosphorus eliminates the coupling. Figure 2 shows the sig-

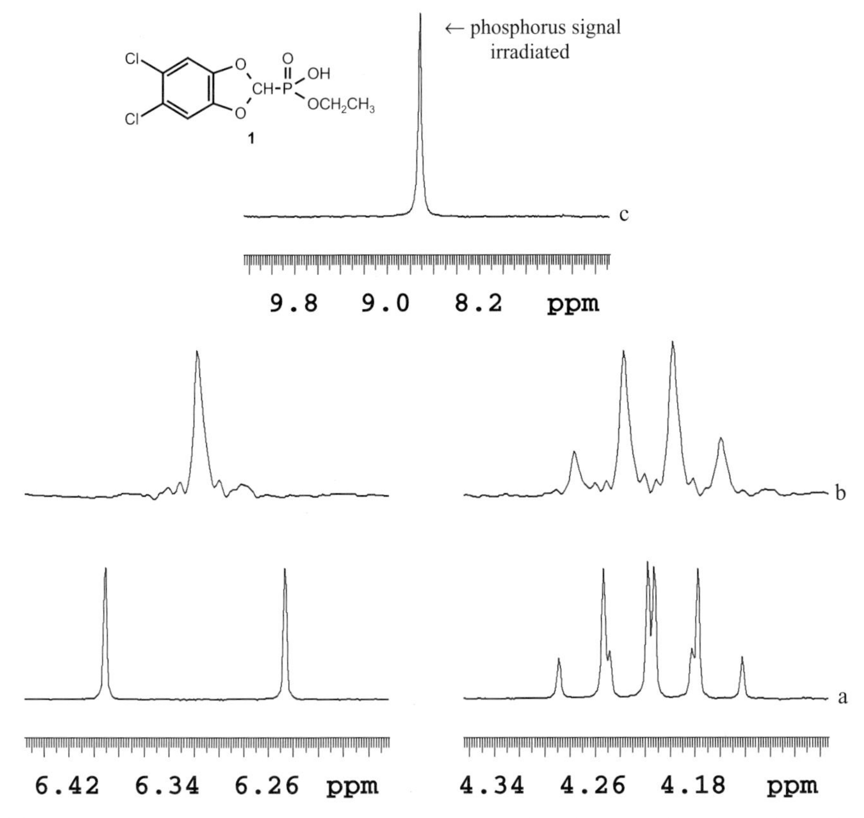

Fig. 2a–c Heterodecoupling experiment on compound **1** (at 200 MHz). **a** undecoupled methine and methylene signals; **b** signals after decoupling of the phosphorus; **c** ^{31}P spectrum, showing the signal which is irradiated using the decoupling channel (channel 2)

nals due to the CH proton (ca. 6.3 ppm) and the OCH_2 protons (ca. 4.2 ppm) before (lower traces) and after (upper traces) decoupling. The top trace shows the ^{31}P signal which is irradiated. On irradiation, the methine doublet is transformed to a singlet, the chemical shift of which lies exactly at the centre of the initial doublet.

The OCH_2 signal at ca. 4.2 ppm in the undecoupled spectrum consists of 8 lines and is due to those methylene protons which have only one oxygen atom in their neighbourhood rather than two. Heterodecoupling reduces the number of lines to 4; we now have a quartet with line intensities 1:3:3:1; thus phosphorus couples with these methylene protons across 3 bonds ($^3J_{P\text{-}O\text{-}C\text{-}H}$). The quartet in the decoupled spectrum (upper trace) is due to coupling of the CH_2 protons with the three equivalent CH_3 protons ($^3J_{H\text{-}C\text{-}C\text{-}H}$): this can be demonstrated by a homodecoupling experiment, a further 2-channel experiment where the second channel is used for *selective* irradiation of the methyl proton signal (a triplet, intensity 1:2:1) at 1.33 ppm (the only signal we have not yet discussed). The

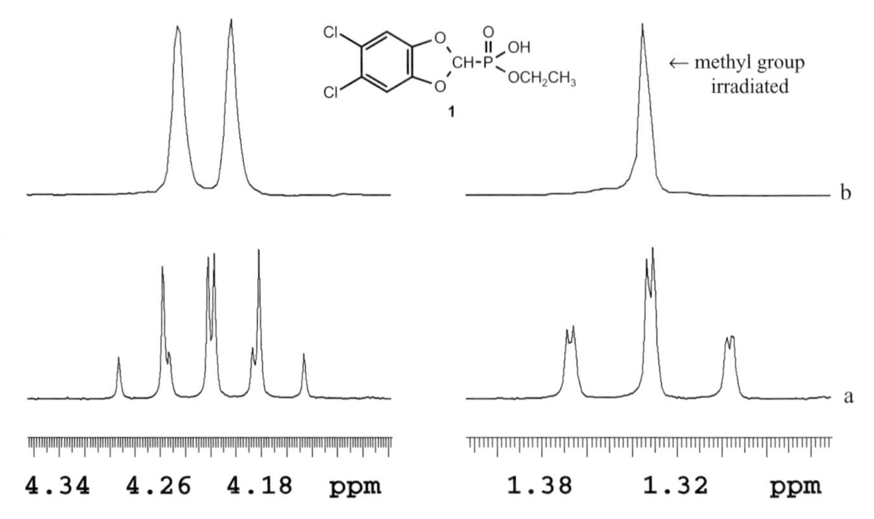

Fig. 3a,b Homodecoupling experiment on compound **1** (at 200 MHz). **a** undecoupled methylene and methyl signals; **b** signals after irradiation of the methyl group

result is now the elimination of ($^3J_{H\text{-}C\text{-}C\text{-}H}$). leading to a doublet signal, the distance between the lines being equal to ($^3J_{P\text{-}O\text{-}C\text{-}H}$).

Thus the original 8-line multiplet is a doublet of quartets (dq).

We can now use a homodecoupling experiment to show that in the methyl signal (triplet, with each line split into a doublet) at 1.33 ppm, the distances between lines 1 and 3, 2 and 4, 3 and 5 or 4 and 6 are equal to ($^3J_{H\text{-}C\text{-}C\text{-}H}$): we irradiate the methylene protons and observe the methyl protons. The result of this experiment is shown in Fig. 3.

There we see the signals due to O*CH*$_2$CH$_3$ on the left and OCH$_2$-*CH*$_3$ on the right. After decoupling (above), the 8-line O*CH*$_2$CH$_3$ signal becomes a doublet due to the P-H coupling, which is of course still present. The 6-line OCH$_2$-*CH*$_3$ signal, the one which is irradiated, becomes one single line. This experiment was carried out on a state-of-the-art spectrometer: earlier spectrometers would more likely have shown the decoupled OCH$_2$-*CH*$_3$ signal in a highly-distorted form.

Homo- and heterodecoupling experiments such as those described here are used routinely in structural analysis and can be carried out very rapidly. In the present case they have provided exact proof that the signal assignments were correct.

1.1.2
Field Dependence of the Spectrum of 1

The decoupling experiments which we have just discussed showed that the multiplet (doublet of quartets) due to the OCH$_2$ group arises from the presence of

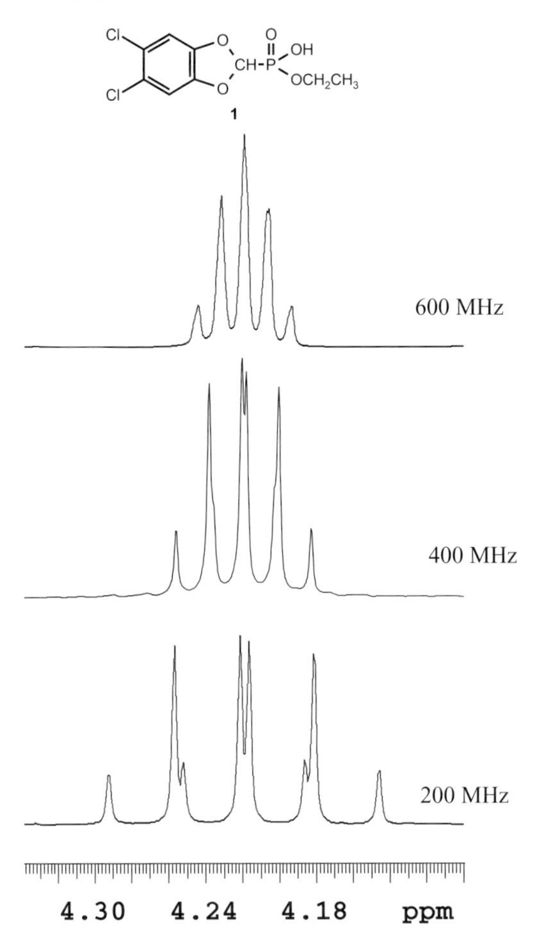

Fig. 4 OCH_2 proton signal of compound **1**, measured using 200, 400 and 600 MHz spectrometers

two coupling constants which are of similar magnitude ($^3J_{HH}$ 7.1 and $^3J_{POCH}$ 8.0 Hz). We could see all 8 lines clearly in the spectrum, which was measured at 200 MHz. If we compare this multiplet with the corresponding signals recorded at 400 and 600 MHz (Fig. 4) we do not see the eight lines so clearly:

This is easy to understand, if we remember that 1 ppm on the chemical shift axis corresponds to 200, 400 and 600 Hz respectively for the three spectrometers. Thus at higher field the multiplet appears "compressed".

Though high fields are invaluable for solving the structures of complex biomolecules, we have found that instruments operating at 200-300 MHz are often in fact better when we are dealing with small molecules.

1.1.3
FID Manipulation: FT, EM, Sine Bell (CH_2 signal of 1)

The signal (FID, free induction decay) resulting from an NMR experiment contains the original data which are stored in the computer, and after the **Fourier transformation** (FT) we obtain the NMR spectrum itself.

We can manipulate the FID mathematically in various ways *before* Fourier transformation, in order to optimise the spectrum with respect to the **linewidth** or the **lineshape.**

Figure 5 shows the original FID and the result when this is multiplied by mathematical functions: either **exponential multiplication** (EM) or **shaped sine bell** (SSB, a sinus function).

EM affects the line width and is often also known as a **line broadening** function LB. A positive value of LB (here 0.8 and 1.9 Hz) broadens the lines, a nega-

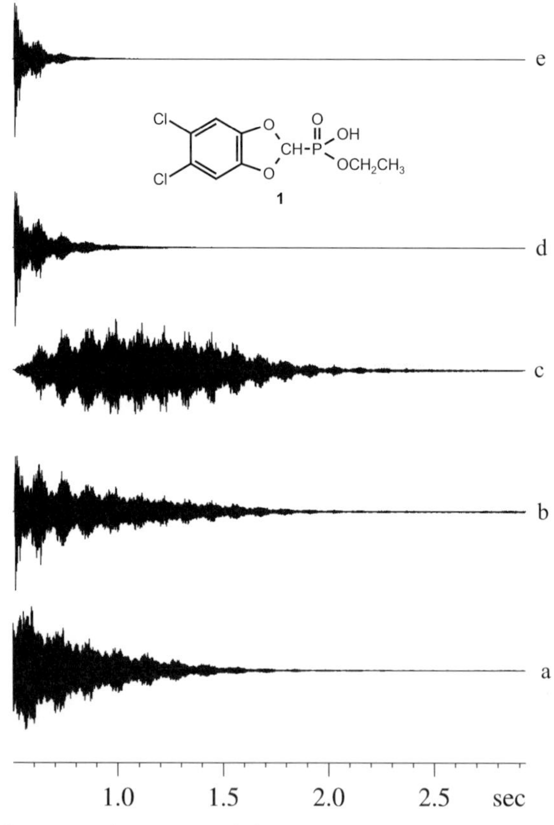

Fig. 5a–e FID of compound **1**. **a** original data; **b** multiplied by a negative line broadening function (–0.3 Hz); **c** multiplied by a shaped sine bell function (SSB = 1); **d** multiplied by a positive line broadening function (0.8 Hz); **e** multiplied by a positive line broadening function (1.9 Hz)

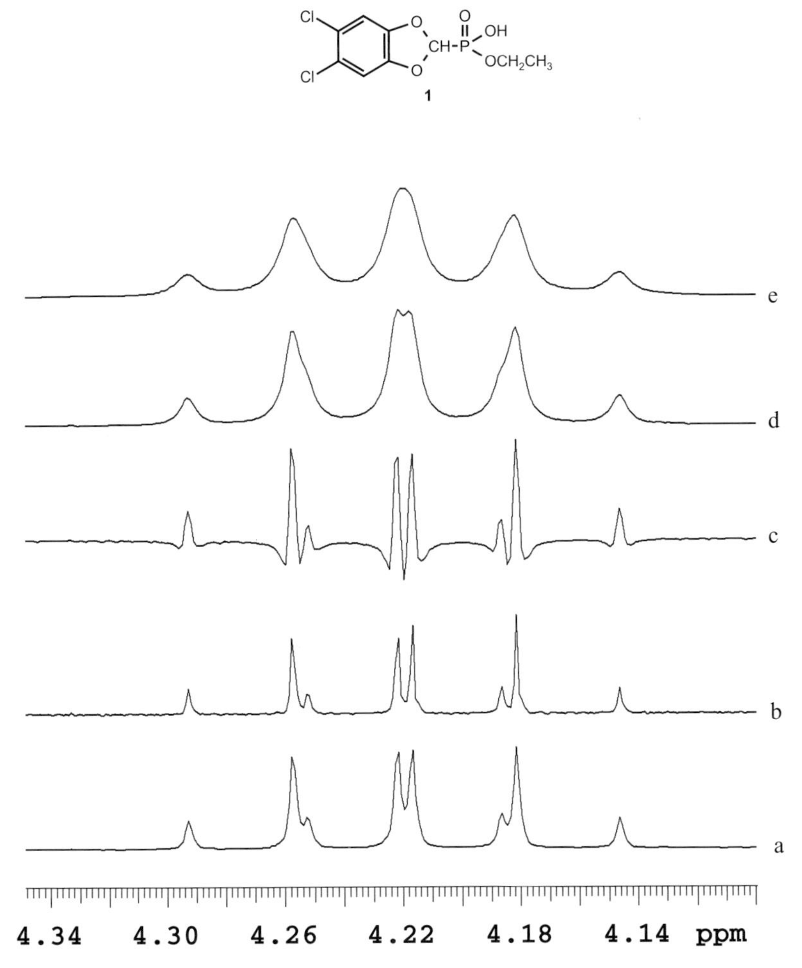

Fig. 6a–e OCH_2 signal of compound **1**(200 MHz). **a** only Fourier transformation; **b** Fourier transformation preceded by multiplication of FID by a negative line broadening function (–0.3 Hz); **c** Fourier transformation preceded by multiplication of FID by a shaped sine bell function (SSB = 1); **d** Fourier transformation preceded by multiplication of FID by a positive line broadening function (0.8 Hz); **e** Fourier transformation preceded by multiplication of FID by a positive line broadening function (1.9 Hz)

tive value (here –0.3 Hz) sharpens them: however, never forget that we are only *modifying* the information present, so that a decrease in the line width is automatically accompanied by an increase in the **baseline noise.** This becomes clear immediately when we see the spectra of the OCH_2 multiplet shown in Fig. 6.

Fourier transformation without data manipulation leads to the multiplet at the bottom (Fig. 6a), which shows more fine structure when a negative LB value is used (Fig. 6b). The spectrum in the middle (Fig. 6c) results from use of the

SSB function, and now all eight lines are clearly visible as the linewidth is much smaller. The price we pay is that the lineshape is completely changed, the positive central "real" lines being accompanied by negative "wings". Positive line broadening functions decrease the quality of the spectra considerably, but there is an improvement of the **signal to noise ratio** (Fig. 6d, e).

The use of sine or cosine functions in FID data processing is an essential tool in 2D NMR.

1.1.4
The proton spectrum of 1 in D_2O or H_2O/D_2O mixtures

The spectra we have so far discussed were recorded using $CDCl_3$, the best all-round solvent for organic molecules. However, many molecules, especially biomolecules, are only soluble in water; biological systems often remain stable only in aqueous solution. Thus NMR measurements in water are extremely important: our model compound is also water-soluble, so that we can use it to demonstrate some important experiments.

We have already mentioned that by simply adding deuterated water to the chloroform solution and shaking the NMR tube leads to H-D exchange, so that the OH signal disappears.

Figure 7 shows the ^1H spectra of **1** dissolved in $CDCl_3$, D_2O, and a 1:1 mixture of H_2O and D_2O.

When we compare Fig. 7a and Fig. 7b we can see that the solvent has an effect on the chemical shift values; such an effect can always occur when the solvent is changed!

The "solvent effect" is due to the interaction between the solute and solvent molecules. D_2O is considerably more polar than $CDCl_3$, so that it can for example interact with the P=O group or the OH group; these interactions influence the neighbouring atoms, so that changes in the chemical shift occur.

In Fig. 7b we observe another very important phenomenon, which can however have unpleasant consequences: the H_2O/HOD signal at 4.7 ppm. D_2O is hygroscopic, so that it should really always be stored in an inert atmosphere. (It is useful to run a proton spectrum of the D_2O in use from time to time to see whether it has taken up water).

If the solute concentration is very low, this signal can become very strong; investigations on biological systems are often carried out in 1:1 mixtures of H_2O and D_2O, and Fig. 7c shows that if we do this for our model compound we see no signal from the dissolved molecules!

There are of course methods for eliminating (or at least partially eliminating) water signals; in fact there are many such methods, and we will demonstrate the use of the simplest of these (which is quite effective), the so-called **presaturation** method. Before carrying out this experiment we need to determine the exact chemical shift of the water signal which we wish to suppress using a standard proton experiment (the computer software can help us here).

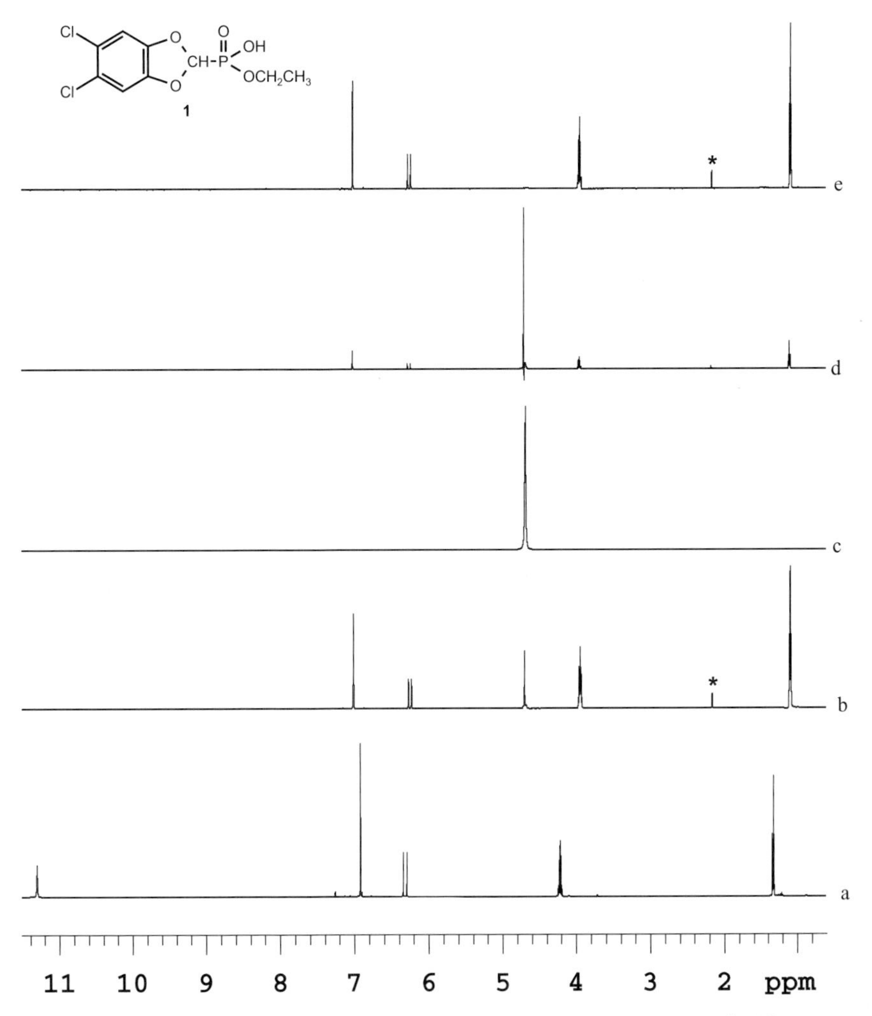

Fig. 7a–e Proton spectra of **1**. **a** dissolved in $CDCl_3$; **b** in D_2O; **c** in D_2O/H_2O; **d** with presaturation of the water signal; **e** with presaturation using a **digital filter**. Signals marked with a * are due to an impurity (solvent from recrystallisation of **1**)

Now comes the actual presaturation experiment, in which the water signal is irradiated for 1–2 s using a pulse set to its chemical shift. This saturates the signal, which is thus no longer visible when the pulse is switched off, and only slowly regains its natural magnitude via **relaxation**. (We shall return to relaxation later).

Now we use a normal proton pulse to excite the solute molecule; Fig. 7d shows the result of the presaturation experiment carried out on the H_2O/D_2O solution of model compound **1**. A residual H_2O/HOD signal can be observed as well as a signal due to the presaturation, but the signals of **1** can be readily seen.

We can improve the appearance of the spectrum by applying a so-called **digital filter**; the result is shown in Fig. 7e.

One thing we can *not* prevent when carrying out presaturation or other water suppression experiments is the distortion or disappearance of solute signals which are very close to (within a few Hz of) the HOD signal!

1.1.5
Integration: Relaxation, T_1, 90°-Pulse, Ernst Angle

So far we have dealt with the chemical shift and coupling constant information in the proton spectrum. What we have not considered is the third important parameter, the signal intensity; this forms the vertical axis of the spectrum, but is not scaled since we do not use intensity units.

The signal intensity gives us quantitative information regarding the individual signals (singlets or multiplets), but this information is only approximate as what we really have to determine are *signal areas*, and the line widths of individual signals can vary considerably.

If we carry out our experiment correctly, the areas of the individual signals are directly proportional to the (relative) numbers of protons giving rise to the signals. As we mentioned under Section 1.1.1, it is advisable to use a pulse angle of 30–40° (the Ernst angle). The integration is carried out by the computer software, and we only need to press the right button or type in the right command in order to obtain the integration curves, which we can also scale with respect to any signal we choose.

Figure 8 shows the result of the integration procedure for compound **1**: it can be presented either as a curve above the signal concerned or, as in this case, as a series of numerical values under the spectrum. Even in the case of pure compounds the integration values are not perfect, but the errors are so small that the *ratio* of the numbers of protons can be easily determined; these numbers are extremely helpful in the structure determination process. Thus here the relative numbers of protons present are given above the individual signals, while the integration values (set with respect to the aromatic protons) are given below the spectrum.

The question arises as to why the integration values are not completely accurate. One reason may be that some multiplets present are too close together, so that the software cannot find the baseline between them. However, there are also systematic errors involved, and this has to do with the **relaxation** phenomenon we mentioned above.

At the very beginning of our discussion in Section 1.1.1, we mentioned that any **pulse experiment** begins with a delay period. This is necessary so that the spins can return to equilibrium before they are excited. After excitation (when the pulse is turned off) we observe the FID, the free induction decay. What

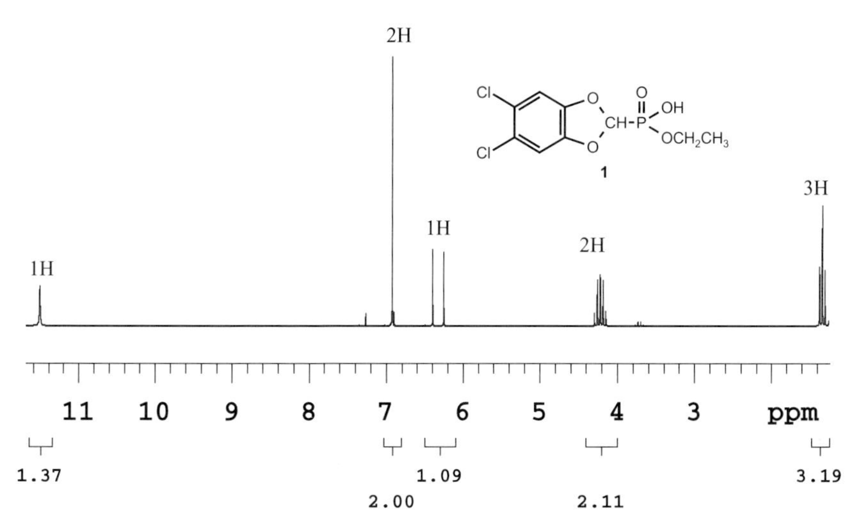

Fig. 8 Proton spectrum of **1** with (below) integration values and (above) numbers of chemically shifted protons in the molecule. The singlet due to the aromatic protons has been set equal to 2 protons

"decays"? The induced magnetisation of the spins, and this process is known as relaxation. It may be slow or fast, as we shall see, and can also occur via a number of processes, which are discussed in detail in the monographs we have recommended for further reading. We will only treat relaxation very briefly here.

We stated previously that the signal induced by a single pulse is largest if we use a so-called **90° pulse**. When the 90° pulse is switched off, the spins "relax", and the time they need to return to equilibrium is obviously longer than if we use a shorter pulse. But a shorter pulse gives us less signal, and so the Ernst angle is a compromise. The time the spins need to return to equilibrium is called the relaxation time, and what we need to talk about here is the so-called **spin-lattice relaxation time T_1** (we are dealing with liquids here, not crystals, and the term "lattice" refers to the local environment of the spins.

In order to design our experiment properly we need to have some idea of how long this T_1 is; relaxation is in fact an exponential process.

T_1 values can be easily determined using **pulse sequences** which form part of the standard computer software, the most common one being the so-called **inversion-recovery experiment.**

This experiment uses two pulses, 180° and 90°, separated by a delay time which is varied. For each delay a certain number of FIDs are accumulated; the result is a series of spectra in which the individual signals have different intensities. Figure 9 shows the result of an inversion-recovery experiment carried out on **1**.

We can see at once that each proton behaves differently, because it has its individual relaxation time T_1; depending on the delay signals may be negative,

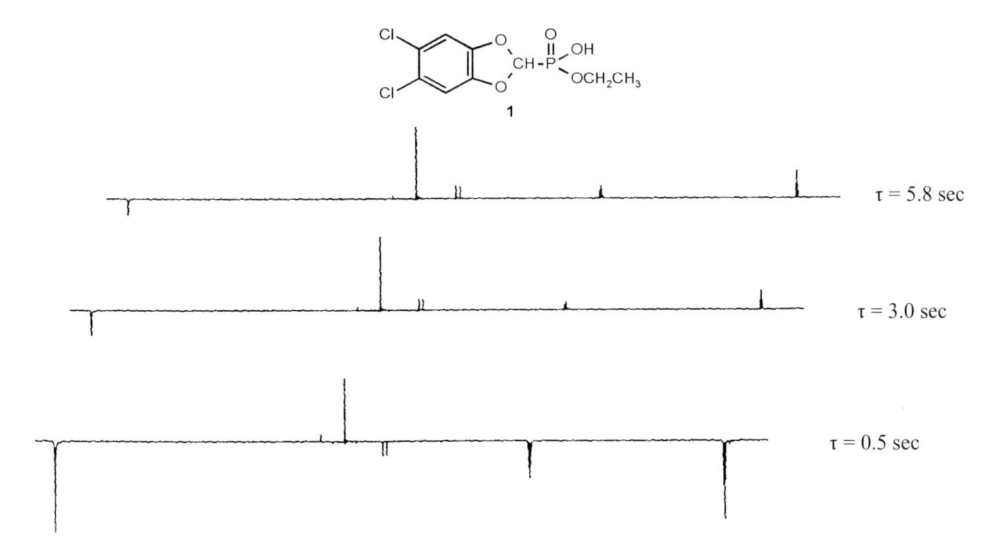

Fig. 9 Spectra of compound **1** obtained from an inversion-recovery T_1 experiment. Pulse sequence: fixed delay – 180° pulse – variable delay – 90° pulse – acquisition of FID

Table 2 Relaxation times T_1 for the protons in compound **1** at 26°C

Spectrometer frequency (MHz)	300	400	500
	T_1 (s)	T_1 (s)	T_1 (s)
OH	0.4	0.5	0.6
Aromatic H	5.4	5.3	5.6
CH	3.3	3.3	3.6
OCH_2	2.7	2.8	2.9
CH_3	2.9	2.8	3.0

positive, or have zero intensity. The T_1 values can be computed using spectrometer software.

One of the textbooks in our list of recommended reading states that proton T_1 values in high-resolution NMR lie close to 1 second and vary little with the type of proton.

We have carried out T_1 measurements for model compound **1** at three different frequencies (300, 400, 500 MHz). The result is shown in Fig. 10 and Table 2.

Our data show that the T_1 values are generally larger than 1 second and vary drastically from signal to signal; they do not appear to vary systematically with the spectrometer magnetic field.

Since the integration values form such an important element of structure determination, we need to set the spectrometer up properly before carrying

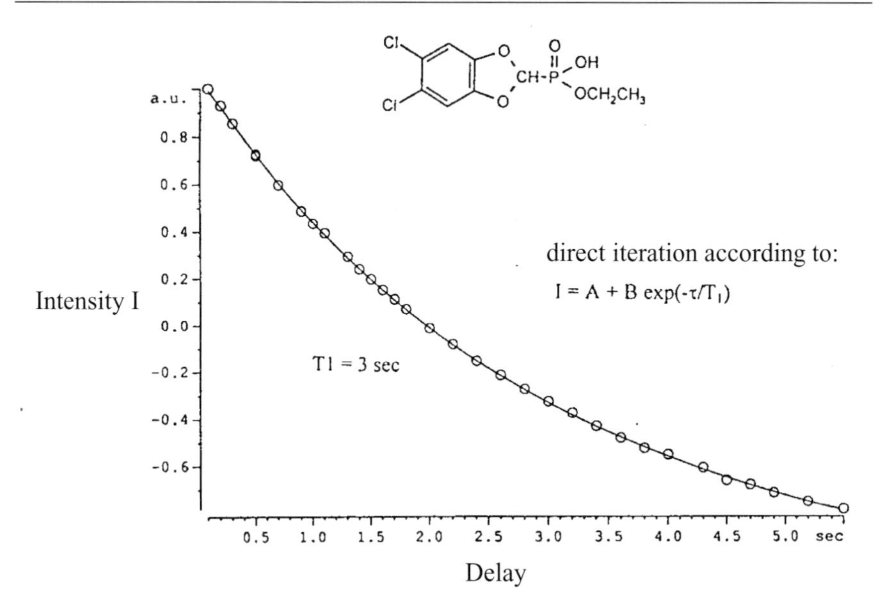

Fig. 10 Compound **1**: T_1 determination for the methyl signal (at 500 MHz in $CDCl_3$ at 26°C). Plot of signal intensity against delay . The computer software gives a T_1 value of 3 seconds

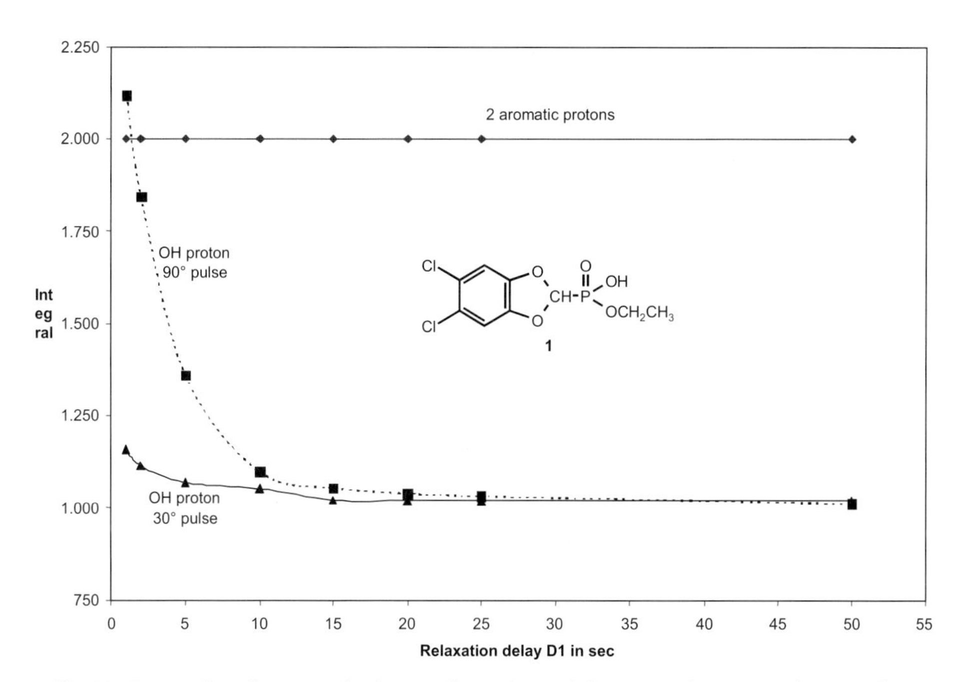

Fig. 11 Comparison between the integration values of the aromatic protons (set equal to 2) and of the OH proton for 90° pulses and 30° pulses as a function of the relaxation delay D1 in seconds

out the NMR experiment. And one very important parameter which is often forgotten is the **relaxation delay**, the delay between the single NMR experiments which allows the nuclei to relax. Remember that relaxation is an exponential process, so that theory suggests that it is necessary for the best results to set this equal to at least *five times* T_1 (in our case more than 25 s for the aromatic protons!). The other parameter we need to set correctly is of course the pulse angle, and the following set of experiments show how these are interrelated.

We carried out two sets of experiments in which we set the pulse angle first at 90°, then at 30°. Using these two values we then varied the relaxation delay. Since the greatest difference in the relaxation times is that between the OH proton and the aromatic protons, we show in Fig. 11 the comparison between the integration values of the aromatic protons (set equal to 2.0) and of the OH proton for 90° pulses and for 30° pulses. The values approach each other with a relaxation delay of 10 s and are virtually equal for a delay of 25 s, but the 90° pulses give values which are completely wrong if a "conventional" delay of 1-2 seconds is used! On the other hand, the error is quite low if the delay is set at 2 s and the pulse length is 30°.

1.1.6
The NOE: Through-Space Interactions between Protons

NOE stands for **Nuclear Overhauser Effect**. Probably only physicists understand the NOE fully, and we shall not go into the theory but only present the results. It is a phenomenon which is useful and important in the NMR of small and large molecules.

We have already seen the result of the interactions between chemically (or magnetically) different protons, the signals from which can be split into multiplets if there is a measurable coupling constant J between them. These coupling constants are the result of the so-called **scalar coupling** in which information about spin states is transferred via the bonding electrons and can be observed across several bonds, depending on the hybridisation of the intermediate carbon atoms. (There is also a so-called **through-space coupling**, but this is not often observed, so that we shall not go into it here).

The NOE depends on another type of coupling, the **dipolar coupling**. This is a strong interaction, but normally we do not notice its effects because the molecules are tumbling rapidly in solution. The magnitude of the dipolar coupling depends on the distance between nuclei, and this is what makes it useful.

We have used signal irradiation to decouple multiplets: this is the phenomenon known as decoupling (see Section 1.1.1). The through-space interactions can also be demonstrated by using signal irradiation, and just as in decoupling we set up the spectrometer so that just one particular proton signal is affected. When we irradiate this signal, we are of course feeding energy into the spin system, thus displacing it from equilibrium: the system tries to get back to equilibrium by using relaxation processes involving the dipolar coupling, and the visible result is changes in signal intensity. These can be positive or negative, depending on (among other things) the size of the molecule: for small molecules they are positive, and the change of the signal intensity is known as the NOE.

These remarks only apply to the proton-proton NOE; experiments involving an NOE between the proton and another nucleus can also be carried out, and the NOE also has an effect on certain carbon-13 spectra, as we shall see later.

Theory tells us that the maximum gain in proton signal intensity is 50%, but normally we are dealing with changes of only a few percent, and the magnitude of these are dependent on the distance between the irradiated proton(s) and the observed ones; the effect is too small to be visible when this distance exceeds about 5Å.

Why is the NOE so important to the NMR spectroscopist? Because it allows us to obtain information about the 3-dimensional structure of the molecule under consideration *in solution* (remember: the only other way to do this is by X-ray structural analysis, but this only works for substances which give good-quality crystals, and by definition not for liquids). Thus we can obtain information on conformations or configurations, something which is particularly important for biomolecules such as proteins, where NOE measurements are absolutely vital.

There are two-dimensional NOE experiments (see below, Section 2.3), but first we shall consider the one-dimensional measurements, which are of two types. To make these clear we shall use molecules **1** and **3**.

1.1.6.1
NOE Difference Spectroscopy

Here we record two proton spectra alternately, one the normal one and the other that in which we irradiate one of the signals. The first spectrum contains no NOE information, while the second does. The resulting FIDs are subtracted from one another by the computer, and the result is a spectrum in which only those signals are present for which intensity differences are observable.

Figure 12c shows such an NOE difference spectrum for the acetal **3**; the spectrum was obtained by irradiating the methine doublet at about 5.8 ppm (the normal spectrum of **3** is shown in Fig. 12a).

A strong negative signal is always observed at the irradiation position. The baseline of the spectrum is very uneven, and it is not possible to correct the **phase** of all the signals at the same time: this is typical of NOE difference spectra, and is due to inexact subtraction of the FIDs. However, we can see a strong positive signal for one half of the **AA′BB′** multiplet due to the para-substituted aromatic moiety: this must be due to the protons closer to the methine proton. No further useful information is available from this experiment, which we can compare with the second technique described below.

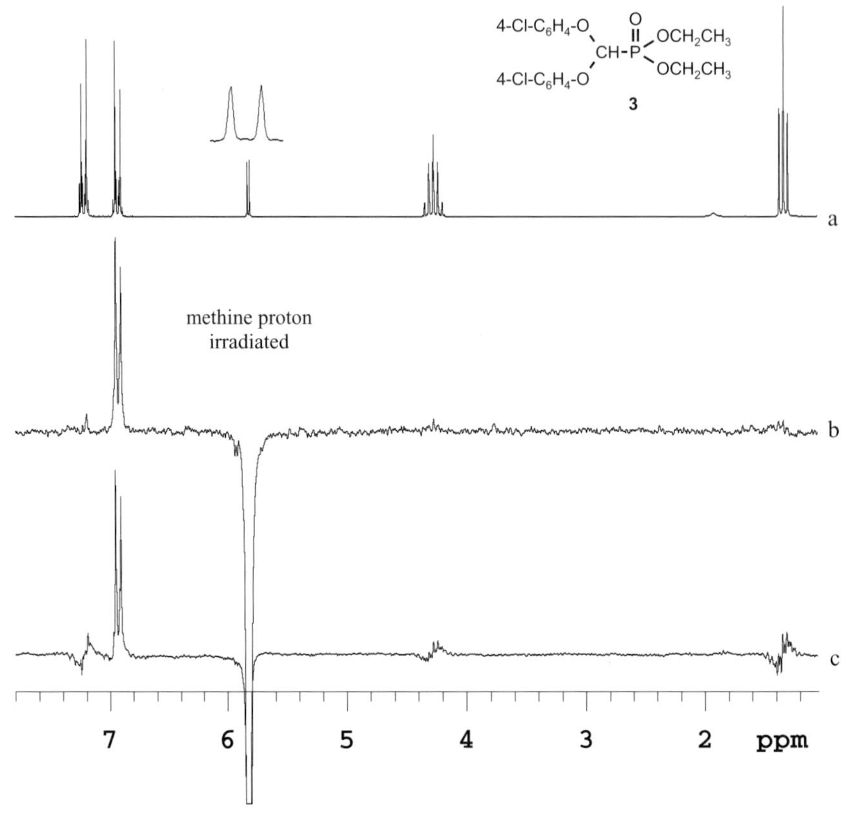

Fig. 12a–c NOE experiments carried out at 200 MHz on compound **3**. **a** normal spectrum, with expansion of methine doublet; **b** selective NOE spectrum, total time required 18 minutes; **c** NOE difference spectrum, total time required (preparation, measurement) 42 minutes

1.1.6.2
Selective 1D NOE Experiment

Advances in computer and spectrometer design have made possible an NOE experiment which does not rely on spectrum subtraction. Again we will not go into details, but this technique relies on excitation of the proton(s) to be irradiated using selective pulses (**shaped pulses** of exactly predetermined width and intensity). The result of such measurements, shown for compound **3** in Fig. 12b, is that only those signals are observed which experience a positive NOE and thus an increase in their intensity. The baseline is now very straight, so that even small signals are clearly visible.

The same proton is irradiated, and just as in the difference experiment, one aromatic pseudo-doublet shows a strong NOE; a very weak but just visible effect is shown by the OCH_2 protons.

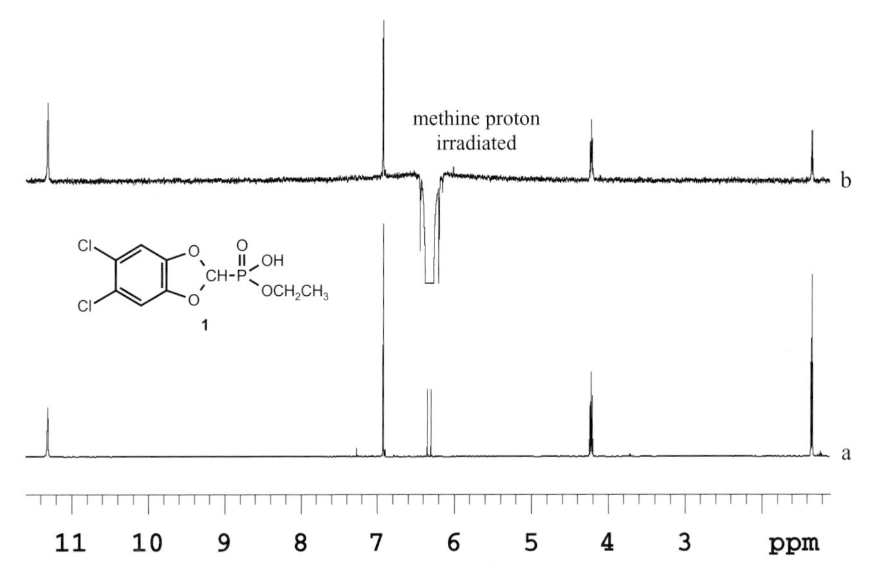

Fig. 13a,b Selective 1D NOE spectrum of **1**. **a** normal spectrum; **b** spectrum recorded with irradiation of the methine doublet (600 MHz, measurement time 4 minutes)

You may wonder why we did not use our model compound **1** in order to demonstrate the NOE. The reason becomes quite clear when we look at the result of a selective NOE experiment carried out at 600 MHz on **1**, which is shown in Fig. 13.

The normal spectrum is shown below, the selective NOE spectrum, again with irradiation of the methine doublet, above.

Although the structural formulae of **1** and **3** are very similar, their NOE behaviour is very different: *all* the protons of **1** show an NOE! The reasons for this become clear when we refer to the known X-ray crystal stuctures of **1** and **3**. Although these depict a defined arrangement in the crystal, whereas NMR spectra reflect averages of possible arrangements in solution, the intramolecular distances measured from the crystal structures do in fact correlate well with the results from the NOE measurements, as is shown in Table 3 below.

Table 3 Distances between the CH proton and other protons in compounds **3** and **1** (in Å)

Compound 3: distance between CH proton and		*Compound 1: distance between CH proton and*	
o-protons	2.07[a]	OH proton	4.04
m-protons	5.93[a]	aromatic protons	4.56[a]
OCH_2 protons	5.02[a]	OCH_2 protons	4.15[a]
CH_3 protons	4.82[a]	CH_3 protons	2.95[a]

[a] shortest distance calculated

In compound **1**, all interproton distances lie in a range which would be expected to give rise to an NOE, as the experiment confirmed. In **3**, although the structural formula is very similar, only the distance between the CH proton and the neighbouring "ortho" protons lies clearly in the "NOE range". The others are close to or above 5 Å, so that only very small NOEs or none at all could be expected.

We have seen that NOE experiments are very useful and can give information on relative interproton distances in the molecule. However, we should stress that NOE experiments can be difficult to interpret because of the many factors involved in their generation.

If and when you need to concern yourself with NOEs in detail, we strongly advise reading up on them in one of the books we recommend in the Appendix.

1.2
^{13}C

Carbon-12, like oxygen-16, is not NMR-active. However, only 1.1% of the total carbon in a molecule consists of the spin-1/2 carbon-13 isotope, so that the sensitivity of this nucleus is much lower. Thus rather than using only perhaps 8 or 16 pulses, as in many proton experiments, we shall now require hundreds or even thousands of pulses, depending on the solute concentration.

1.2.1
Natural Abundance Carbon-13 Spectrum of Compound 1

Organic compounds contain four types of carbon atom: methyl, methylene, methine and quaternary. And so if we simply record the spectrum as we would a proton spectrum, the result will be a series of quartets, triplets, doublets and singlets, each associated with a carbon-proton one-bond coupling constant of between 125 and 250 Hz. If we are dealing with a complex molecule, these multiplets will overlap and give us spectra which are almost impossible to analyse. In addition, coupling interactions over two or more bonds complicate the picture still further.

Thus when it became possible to record carbon-13 spectra routinely it was decided that the logical thing to do would be to decouple ALL of the protons from the carbons simultaneously (a technique known as **broad-band decoupling**) in order to obtain a carbon-13 spectrum consisting only of singlets.

This gives us the chemical shift information for each type of carbon atom in the molecule. We do not have any coupling information, however, but we shall see below how we can obtain the coupling information we need.

Let us look at the natural abundance carbon-13 spectrum of our model compound **1**, which is shown in Fig. 14.

If we count the number of different carbons in the molecule, we see that we expect six signals (three for the aromatic carbons, one for the methine carbon,

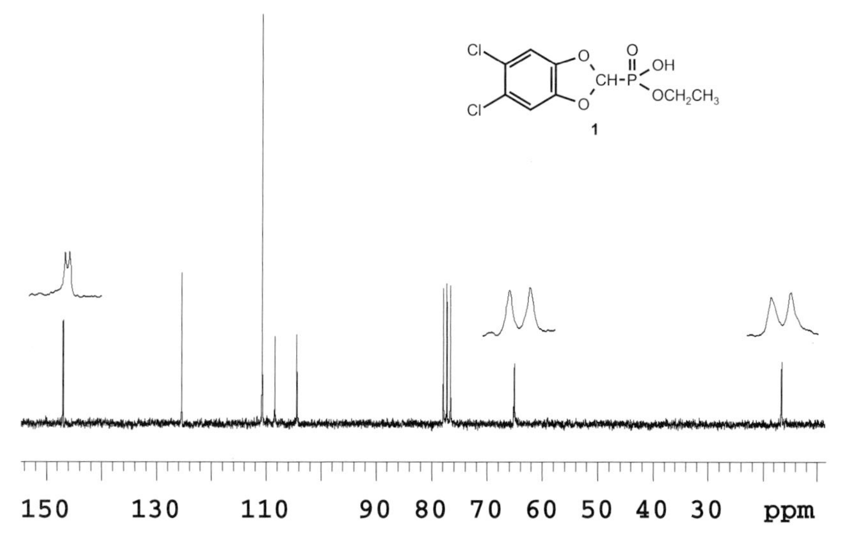

Fig. 14 Natural abundance carbon-13 spectrum of **1** (50 MHz) with expansion where necessary to show doublet structure. The assignments are as follows (from left to right): aromatic C bonded to oxygen (doublet) ; aromatic C bonded to chlorine (singlet); aromatic CH (singlet); methine (doublet); $CDCl_3$; OCH_2 (doublet); CH_3 (doublet). Multiplet splittings are due to coupling with phosphorus and are (except for $^1J_{PC}$) small

one for the methylene and one for the methyl carbon). Each of the three aromatic carbon signals corresponds to two carbon atoms, the other three signals each correspond to one carbon atom. Some of these signals will certainly be split into doublets because of the presence of carbon-phosphorus coupling. We shall also see a signal due to our solvent $CDCl_3$; this absorbs at 77 ppm and is a triplet because of coupling between carbon and deuterium (deuterium being a nucleus with spin $I = 1$).

The rule in carbon-13 NMR is that sp^2-hybridised carbons (carbonyl, aromatic, olefinic) absorb at lowest field, followed by sp-hybridised (acetylenic, nitrile) and sp^3 (aliphatic). A first glance leads us to believe we have seven signals, but we must remember that the methine carbon is directly bonded to phosphorus, so that we shall expect a relatively large C-P coupling. The other C-P couplings will probably be very much smaller.

So the seven signals reduce to six, one obviously being a doublet. If we expand the spectrum we see that another three signals are doublets with a small C-P coupling.

Before we try to assign the signals, let us look at the signal intensities. These are obviously not as we would expect, but are very uneven. There are two reasons for this, one having to do with the NOE and one with relaxation.

We have so far looked at the NOE only in a homonuclear manner, but of course there is also a heteronuclear NOE. Theory tells us that when we are deal-

Table 4 Result of a prediction compared with the actual values

Chemical shift (ppm)	J_{CP} (Hz)	Calculated shift	J_{CP} (calc.)	Assignment
147.1 (d)	2.3	152.0	8.0	C_{arom}-O
125.5 (s)	0	128.4	0	C_{arom}-Cl
110.9 (s)	0	115.5	4.8	C_{arom}-H
106.5 (d)	201.3	102.6	207.2	CH-P
65.2 (d)	7.2	61.9	6.0	OCH_2
16.7 (d)	5.5	15.5	8.0	CH_3

ing with C-H fragments in small molecules, the decoupling of the proton leads to an increase in the carbon signal intensity by up to almost 200%! So signals of protonated carbons should be stronger than those of non-protonated carbons.

Obviously we cannot however simply correlate the signal intensities with the presence of attached protons. So relaxation must also play a very important role. Relaxation times T_1 for carbon atoms also depend on whether these are protonated or not, and while T_1 for methyl or methylene groups may only be a few seconds, it may be as long as around two *minutes* for quaternary carbons! Now the choice of an ideal relaxation delay becomes impossible, and so we have to make compromises, which result in the large variations in signal intensity.

The story is even more complicated than we have suggested, because carbon can relax by more than one mechanism. Protons rely on ***dipole-dipole relaxation****, which also works well for protonated carbons but badly for non-protonated carbons. But carbon also for example makes use of* ***spin-rotation*** *relaxation, which is particularly active for methyl groups. And the magnetic field dependence of the various mechanisms also differs. We realise that relaxation is a very difficult subject, and if you want to know more then there are plenty of textbooks available!*

So basically there is no point in integrating a broad-band decoupled carbon spectrum. This is not so much of a drawback as it sounds, because the signals are distributed over a range of more than 200 ppm, so that line overlap is very unusual.

Signal assignment can be done in several ways: the simplest is to use prediction programmes, and Table 4 presents the result of a prediction compared with the actual values.

As we can see, the predicted chemical shifts and coupling constants agree well with the actual values.

1.2.2
Coupled Spectrum (Gated Decoupling)

The proton-decoupled spectrum (Fig. 14) made it easy for us to assign the signals to the different carbon atoms, particularly because of the help given by the carbon-phosphorus coupling. However, the information which is „lost“ during decoupling, the presence or absence of carbon-*proton* coupling, can be very important in many cases. Thus the degree of s-character in a C-H bond plays an important role in determining the value of $^1J_{CH}$, while the value of $^3J_{CH}$ is very important for solving stereochemical problems, since, like $^3J_{HH}$, it shows a **Karplus**-type dependence on the **dihedral angle** subtended by the C-H and C-C bonds involved.

It is in fact quite simple to record a carbon-13 spectrum with the broad-band decoupling switched off. Such a procedure has the disadvantage that the gain in signal intensity due to the NOE is lost, so that measurement times are very long.

There is however an experiment which allows us to obtain a coupled spectrum *without* losing the NOE effect: this is known as **gated decoupling.** Here the computer has to control some elegant switching in which the broad-band decoupling is ON during the relaxation delay, allowing the NOE to build up. It is however OFF during the pulse and during the acquisition, so that we can still retain the coupling information.

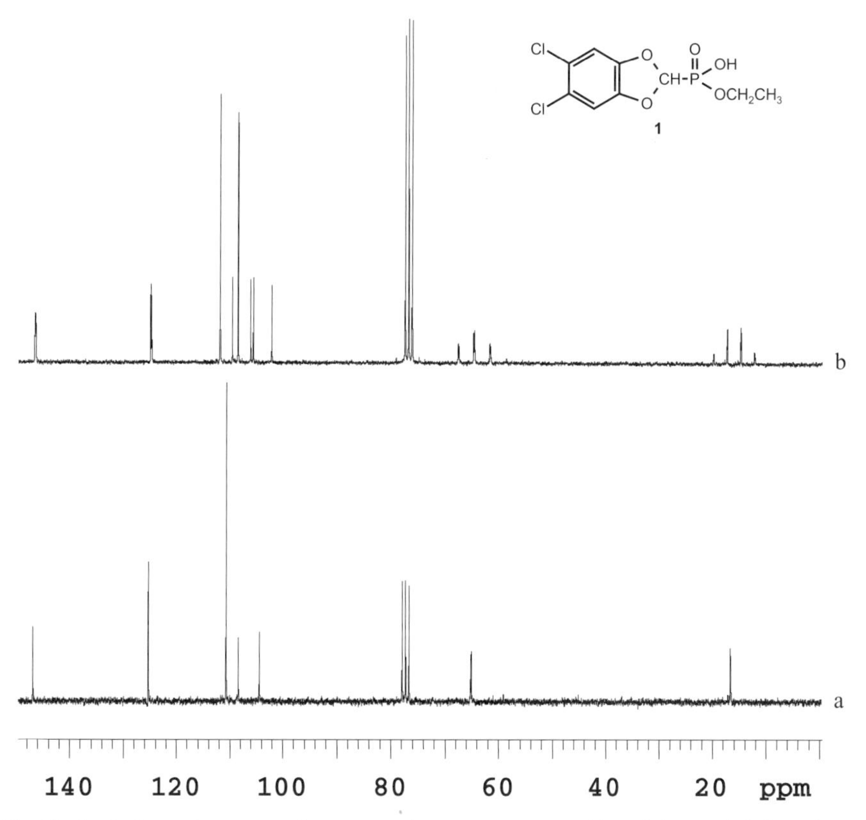

Fig. 15a,b Carbon-13 spectra of compound **1**. **a** proton broad-band decoupled; **b** carbon-proton coupling present (gated decoupling)

Figure 15 shows the normal broad-band decoupled and gated decoupled spectra of compound **1**; in the latter we can see the multiplets arising from C-H coupling (across one or more bonds) and C-P coupling. The rules for the number of lines in a multiplet and their intensities are the same as for protons, since ^{13}C and ^{31}P are both spin-1/2 nuclei.

1.2.3
Quantitative ^{13}C Spectrum (Inverse Gated Decoupling)

Because of the NOE and differences in relaxation rates, the intensity differences for carbon signals in a broad-band decoupled spectrum are extremely large, so that quantitative information is not available.

Though this is generally not a problem, there is an experiment available which allows us to obtain reliable quantitative intensity information, which we may for example need when studying mixtures of compounds.

This experiment, the results of which are shown in Fig. 16, is known as **inverse gated decoupling**: the broad band decoupling is OFF during the relaxation delay, so that no NOE can build up. It is however switched ON during the radio frequency pulse and during the acquisition, so that the C-H coupling is eliminated (the C-P coupling is not affected). Thus, as shown in the upper spectrum, no C-H coupling is present, and the intensities of the carbon signals are correct. The lower spectrum shows the integration values for the standard carbon-13 experiment, which are clearly completely incorrect: in each case the signal on the left is set equal to two (carbons), and while the intensities in the upper spectrum lie within 10% of the true values, most of those in the lower spectrum are too high by factors greater than two.

However, in this experiment it is very important that the relaxation delay chosen is very long, since the carbon atoms have very different relaxation times (and relax by different mechanisms). In our example the relaxation time was set to 120 seconds! This of course makes the experiment a very time-consuming one (28 hours measurement time!).

The integration of the various carbon signals now gives intensity values which are sufficiently accurate for most purposes.

1.2.4
Decoupled Spectrum: Proton Decoupling, Proton and Phosphorus Decoupling

The signals in the coupled carbon-13 spectra are split by the C-H couplings, and the values of J_{CH} can be directly read off. If for example we consider the chlorine-bearing carbons in our model compound **1** (Fig. 17), the resulting signal is split into a doublet of doublets, due to the coupling with the two aromatic protons. The coupling paths are different: we observe both $^{2}J_{CCH}$ and $^{3}J_{CCCH}$, the values being 5.4 and 7.9 Hz respectively.

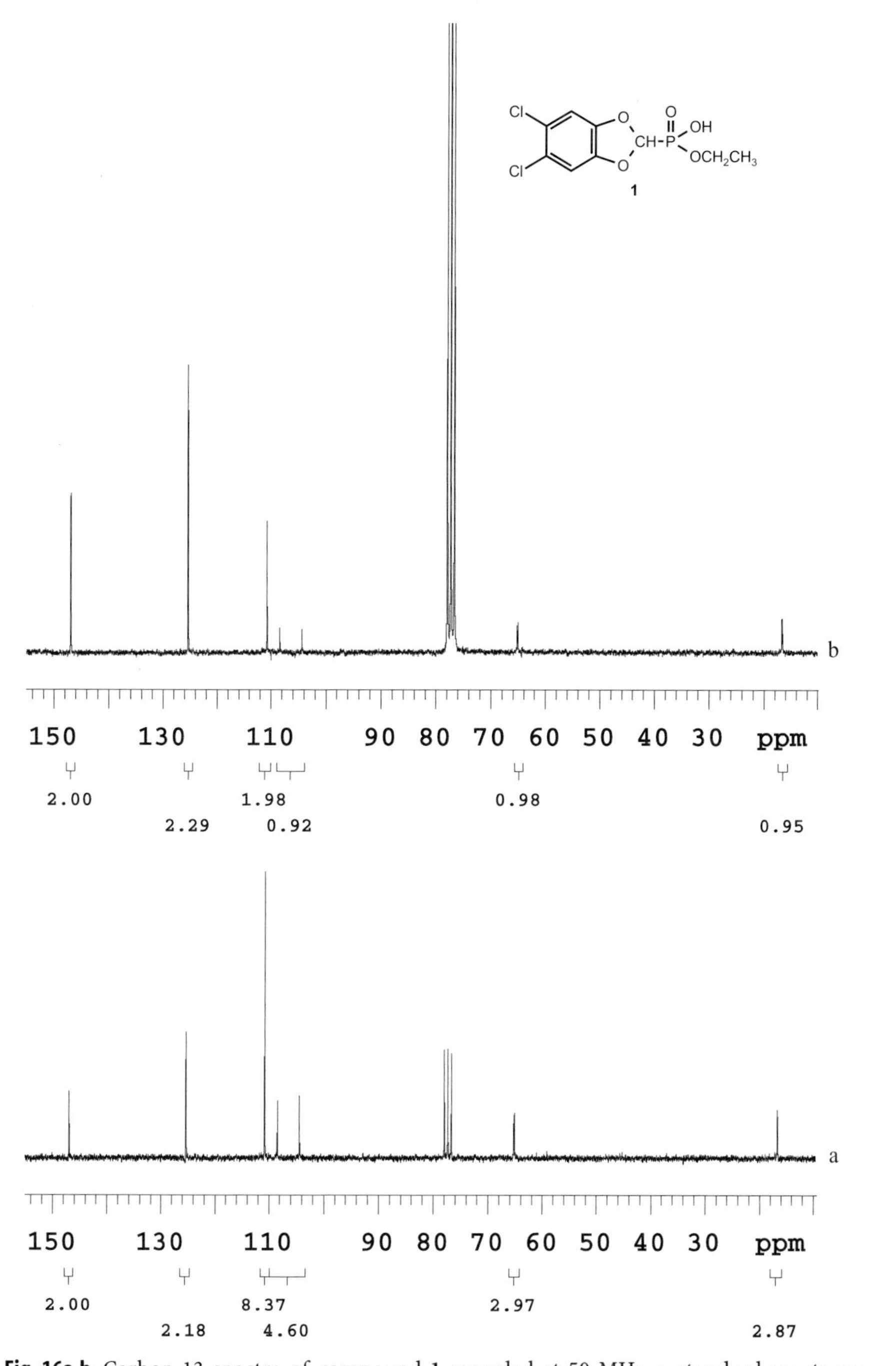

Fig. 16a,b Carbon-13 spectra of compound **1** recorded at 50 MHz. **a** standard spectrum with integral values (measurement time 1.5 hours); **b** inverse gated decoupled spectrum with integral values (measurement time 28 hours!)

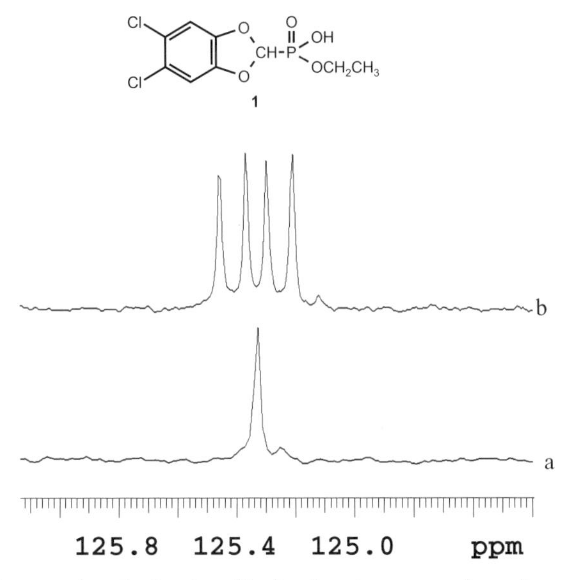

Fig. 17a,b Carbon-13 signals for the chlorine-bearing aromatic carbons in **1**. **a** proton decoupled; **b** no proton decoupling

The determination of the coupling constants is more difficult for other signals. Thus the methyl carbon of **1** (Fig. 18, lower trace) is split into a quartet by the three methyl protons. However, the four lines of the quartet are split further (into doublets of triplets), since the couplings with the P nucleus ($^3J_{POCC}$) and with the two protons of the OCH_2 group ($^2J_{HCC}$) are also readily visible.

The determination of these two coupling constants can be carried out using a **selective proton decoupling** experiment. The middle trace in Fig. 18 shows the results of such an experiment.

Here we have irradiated the OCH_2 group in the proton spectrum: the result is a doublet of quartets with two coupling constants ($^1J_{CH}$ = 127.7 Hz, $^3J_{POCC}$ 5.5 Hz). We can thus extract $^2J_{CCH}$ from the multiplets in Fig. 18b; its value is 2.7 Hz.

For completeness, the upper trace in Fig. 18 shows the broadband-decoupled signal, which is of course just a doublet due to the P-C coupling.

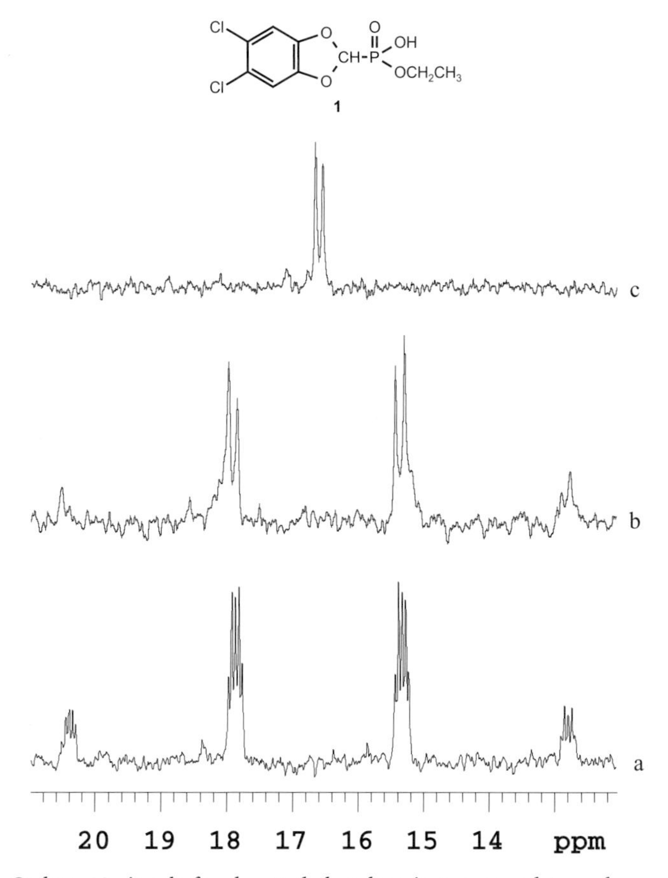

Fig. 18a–c Carbon-13 signals for the methyl carbon in **1**. **a** complete carbon-proton coupling present; **b** selective decoupling of methylene protons; **c** broad-band decoupled

1.2.5 *APT, DEPT, INEPT*

In the carbon-13 experiments so far discussed, only a single radio-frequency pulse has been used to irradiate the spin system. This gave us information on the chemical shifts of the carbon nuclei in the molecule. The coupled spectrum obtained using gated decoupling (Section 1.2.2) told us how many protons are bound to any one carbon atom; however, this experiment requires a lot of time. There are however other experiments which give us this information on the "**multiplicity**" of the carbon atom (quaternary, methine, methylene, methyl) which can be carried out very quickly. Such experiments, which are invaluable in structural determination work, will be discussed here. The two most important are **APT** (Attached Proton Test) and **DEPT** (Distortionless Enhancement by Polarisation Transfer).

Both find their origin in the ***spin-echo sequence****, devised by Hahn in 1952 and used for the determination of relaxation times.*

The theory behind both of these experiments, and in particular the DEPT experiment, is rather complicated, so that we refer you to NMR textbooks for details. The important feature of both is that the carbon signals appear to have been simply broad-band decoupled, but that according to the multiplicity they appear either in positive (normal) phase or in negative phase, according to their multiplicity.

APT distinguishes between two groups of signals, methyl/methine (normally shown in positive phase) and methylene/quaternary (negative). DEPT is similar, except that quaternary carbons are not detected by this sequence. There may be cases where it is necessary to distinguish between methyl and methine, and this can be done by adjusting the DEPT pulse sequence (known as "editing"): the standard experiment is known as DEPT-135, and requires, like APT, short measurement times.

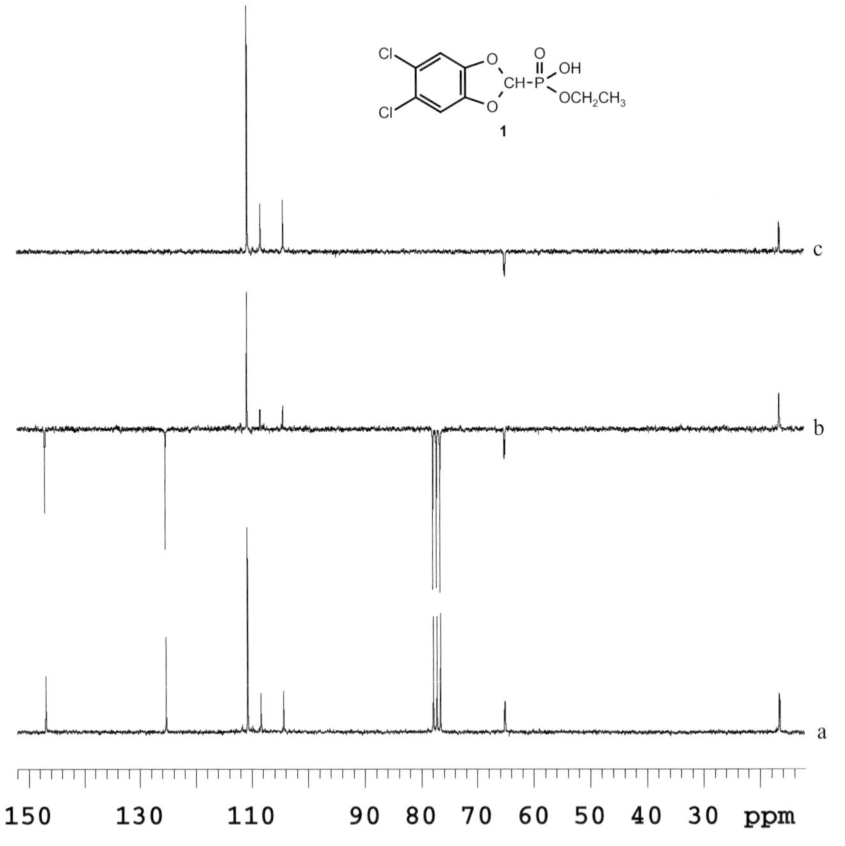

Fig. 19a–c Carbon-13 spectra of compound **1**. **a** standard spectrum (broad band decoupling); **b** APT spectrum; **c** DEPT-135 spectrum

Figure 19 shows the normal (broad-band decoupled), APT and DEPT-135 spectra of model compound **1.** Note that in the APT spectrum the solvent ($CDCl_3$) is visible, but not in the DEPT spectrum, where the two low-field quaternary aromatic carbons are also absent.

There is another member of this family of experiments known as **INEPT** (Insensitive Nuclei Enhancement by Polarisation Transfer), which was the forerunner of DEPT. INEPT still has its uses for obtaining spectra of really insensitive nuclei such as silicon-29 or nitrogen-15.

1.2.6
The INADEQUATE Experiment

The information on carbon chemical shifts and multiplicities is invaluable for structure determination. It would be ideal if we also had a method for obtaining information directly on carbon-carbon bonding in the compound under study, since this would allow us to draw on paper at least parts of the carbon framework of the molecule.

Carbon-13 represents only 1.1% of the total carbon nuclei present in a sample. In order to get the information we require, we need to detect the doublets due to carbon-carbon coupling. Thus we wish to observe only those molecules containing two neighbouring carbon-13 nuclei, i.e. about 10^{-4} of the nuclei present; at the same time we have to get rid of the signal coming from those molecules containing only one carbon-13 (the great majority!).

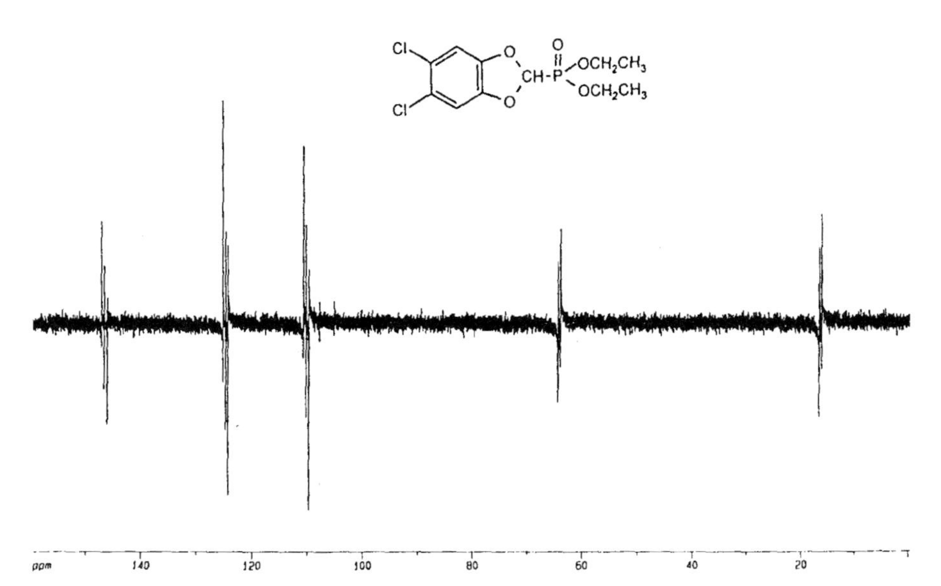

Fig. 20 1D INADEQUATE spectrum of compound **2** (75 MHz, 50% solution in $CDCl_3$, measurement time 20 hours). Note that the multiplets are distorted because they could not be correctly phased

INADEQUATE stands for Incredible Natural Abundance DoublE QUAntum Transfer Experiment. Again, we refer you to NMR textbooks for an explanation of the principles. Here we only present the result, which is shown in Fig. 20 for the diester **2**.

Note that we have used a highly concentrated solution, but even so required 20 hours to obtain the spectrum. This is because there are so few molecules containing the fragment ^{13}C-^{13}C present. At first glance the spectrum looks very strange,

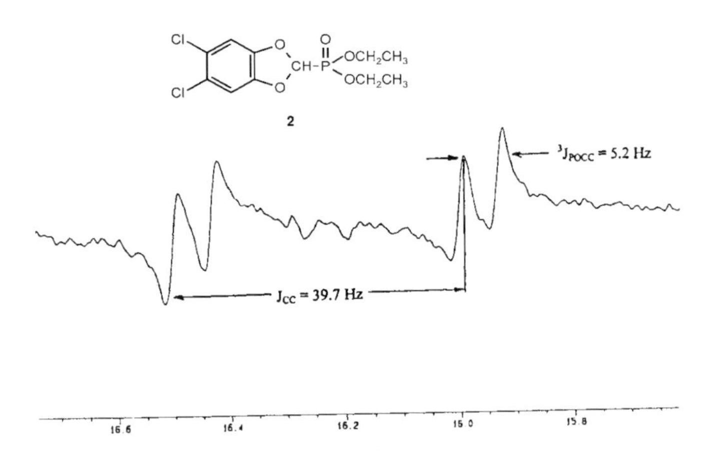

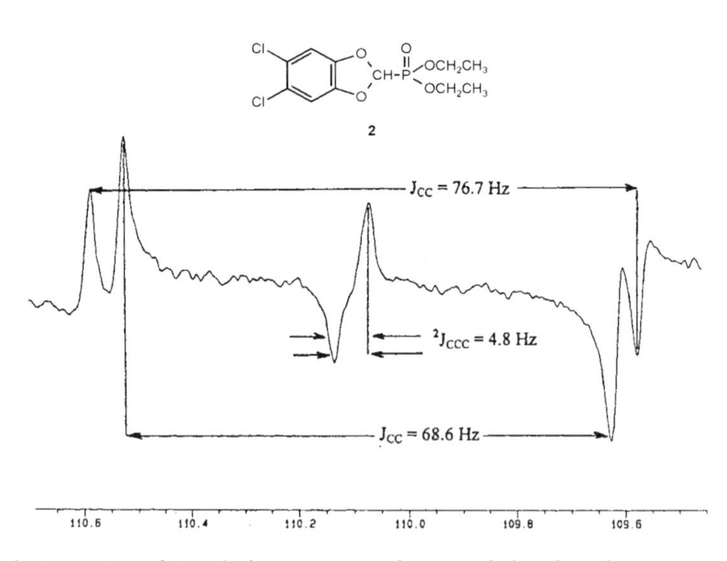

Fig. 21a,b Expansion of signals from compound 2. **a** methyl carbon; **b** aromatic CH carbons

and if we count the number of signals we only find five. However, one carbon in the molecule, the methine carbon bonded to phosphorus, does not have a direct carbon neighbour, so it cannot appear. The other five signals appear at the correct chemical shifts, but they consist of multiplets which are not in phase.

Why do we see multiplets rather than singlets? Firstly, we are in each case looking at signals due to C-C coupling, so each signal will be split into a doublet just as in an **AX proton spin system**. Secondly , the influence of the phosphorus nucleus is still there and will lead to further splitting of some of the signals.

To see clearly what is going on, we need to expand the signals, and this is done in Fig. 21 for the methyl signal at 16.2 ppm (above) as well as for the protonated aromatic carbon at 110.1 ppm (below).

The upper signal consists of a doublet of doublets, with two coupling constants: 39.7 Hz and 5.2 Hz. The first is the one-bond coupling constant $^1J_{CC}$, while the second is $^3J_{POCC}$, which we have already observed in the normal carbon-13 spectrum. The methylene signal would look similar if we expanded it.

The lower signal is more complicated, and before we can interpret it exactly we need some background information. The magnitude of one-bond C-C coupling constants depends on bond hybridisation (ethane 35, ethene 68, benzene 56, ethyne 172 Hz), while two- and three-bond C-C couplings are very small, often around 2-5 Hz. The second thing we have to remember, and this is a new concept, is that the lines in the multiplets from INADEQUATE spectra often come from *different* spin systems!

Table 5 Results from 1D INADEQUATE experiment carried out on diester **2**

Chemical Shift (ppm)	Assignment	Coupling Constants (Hz)
146.6	C_{arom}-O	$^1J_{CC}$ 76.7[a] $^2J_{COC}$ 3.2 $^3J_{CCCC}$ 5.7[b] $^3J_{PCOC}$ 2.2
124.7	C_{arom}-Cl	$^1J_{CC}$ 68.6[a] $^3J_{CCCC}$ 5.2[b]
110.1	C_{arom}-H	$^1J_{CC}$ 76.7 $^1J_{CC}$ 68.6 $^3J_{CCCC}$ 4.8[b]
106.4	CH-P	$^2J_{COC}$ 3.2 $^1J_{CP}$ 197.8
64.0	OCH_2	$^1J_{CC}$ 39.7 $^2J_{COP}$ 6.8
16.2	CH_3	$^1J_{CC}$ 39.7 $^3J_{CCOP}$ 5.2

[a] should show two direct couplings, but these are apparently almost equal
[b] the two-bond coupling $^2J_{CCC}$ is smaller than the three-bond coupling $^3J_{CCCC}$ and causes only line broadening

Thus here we see two large doublet splittings, one between the CH carbon and the CO carbon ($^1J_{CC}$ 76.7 Hz) and one between the CH carbon and the CCl carbon ($^1J_{CC}$ 68.6 Hz).

These are due to two different ^{13}C-^{13}C spin systems. The third ^{13}C-^{13}C spin system leads to a doublet in the centre of the multiplet with a small splitting: this is $^3J_{CCCC}$ and equals 4.8 Hz.

The complete C-C and C-P coupling information is given in Table 5.

1.3
^{31}P

Phosphorus is an unusual element, because it has only one single isotope, phosphorus-31, and that this isotope is NMR-active with a spin of ½. The only other elements for which this is the case are fluorine, yttrium, rhodium and thulium.

The sensitivity of ^{31}P is also high, so that measurements do not require high sample concentrations.

1.3.1
Natural Abundance Phosphorus-31 Spectrum of Compound 6

Since the phosphorus spectra of compounds **1** to **5** are rather boring (only one phosphorus resonance), we shall also use compound **6**, which contains three non-equivalent phosphorus nuclei, to demonstrate the results of the experiments we describe.

1.3.2
Proton-Decoupled and Proton-Coupled Spectra

Compound **1** contains one phosphorus atom, so that the broad-band proton decoupled spectrum is extremely simple: it consists of only one line at 8.5 ppm. This spectrum is shown in Fig. 22, together with the proton-coupled spectrum.

The proton-coupled spectrum is much more informative. We can see immediately which protons show a measurable coupling with the phosphorus atom, because the pattern is clearly identifiable as two triplets separated by 28.7 Hz. This, as we have already seen in Table 1, is the two-bond coupling between the phosphorus and the methine proton. The triplets (intensity 1:2:1) result from the three-bond coupling between the phosphorus and the methylene protons, which is 8.0 Hz. If we look at the linewidths (width at half height, measured in Hz) we can see that in the proton coupled spectra they are larger. This is because there are further non-resolvable couplings present which are smaller than the signal linewidths; this was suggested in the simulation results for the proton spectrum, presented in Section 1.1.1.

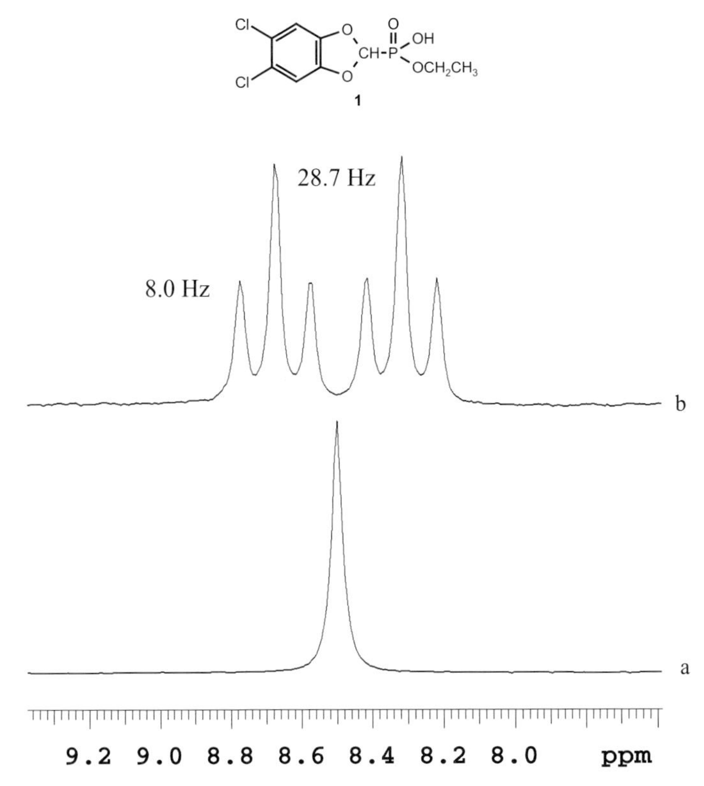

Fig. 22a,b Phosphorus-31 spectra of compound **1**. **a** protons decoupled; **b** proton-phosphorus coupling present

1.3.3
Coupled Spectrum (P-P Coupling)

It is by no means unusual to come across compounds which contain more than one phosphorus atom: Fig. 23 shows the proton decoupled coupled phosphorus spectrum of compound **6**, which contains three chemically different phosphorus nuclei. Phosphorus behaves in NMR just like the proton, so we shall expect to see three signals, split into multiplets if there is an observable coupling between the phosphorus nuclei.

In all cases the oxidation state of phosphorus is five, and the chemical shift range observed is only about 12 ppm. Note that the two phosphorus atoms attached to the methine carbon are non-equivalent because they are chemically different (phosphonate and phosphine oxide). We can expect the coupling between ^{a}P and ^{b}P to be large, as they are separated by two bonds, while that of ^{a}P to ^{b}P or ^{c}P will be small (coupling over five bonds).

We have labelled the three signals, two of which are additionally shown in an expanded form, and it is clear that the low-field signal, with the coupling of 9.4

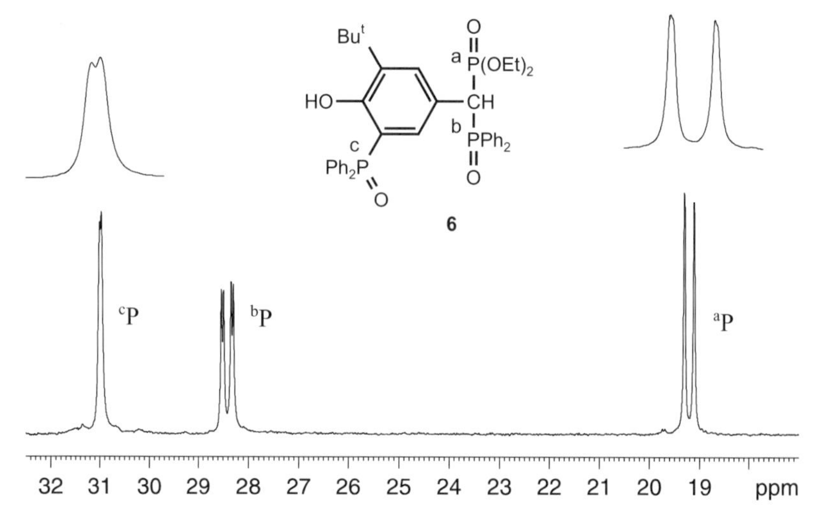

Fig. 23 Phosphorus-31 spectrum (202 MHz) of compound **6**, measurement time 2 minutes

Hz, must correspond to ^{c}P. We can see only *one* other coupling (39 Hz), which occurs in both of the other multiplets: this must be between ^{a}P and ^{b}P. But which signal corresponds to ^{a}P and which to ^{b}P? This information comes from the chemical shift of analogous compounds, where phosphonates such as ^{a}P absorb to high-field of 20 ppm, while phosphine oxides such as ^{b}P and ^{c}P absorb at around 30 ppm.

But we have a puzzle here: since rotation around the aryl-CHP_2 bond should be relatively unhindered, why does ^{c}P not couple to *both* ^{a}P and ^{b}P? We will return to this question when we discuss the 2D phosphorus-phosphorus correlation experiment.

2
2D Experiments

2.1
General Principles, Inverse Techniques, Gradients

Two-dimensional NMR? A strange concept, when we consider that all the spectra we have previously dealt with were of course plotted in two dimensions, the two axes (dimensions) being a frequency axis (horizontal, expressed in ppm rather than in Hz for reasons we have already discussed) and an intensity axis. To understand the basic idea of two-dimensional NMR (**2D NMR**) we should first remind ourselves that while the spectrum we see and use is plotted as a function of frequency, it was originally recorded (as the FID) as a function of time. Only after the Fourier transformation did it become intelligible to us.

So the "one dimension" in the previous spectra was a time dimension, and to extend NMR to two dimensions involves recording the spectrum as a function of two time variables (time dimensions) and carrying out a **double Fourier transformation** to give us an understandable spectrum. This also naturally contains intensity information, providing us with information in three dimensions. But the intensity information is less useful, so we choose a representation of the spectrum which is called a "contour plot", basically similar to the way maps can be drawn with contour lines showing the heights above sea level. Since liquid samples normally give sharp lines, our "mountains" are more like needles and their contour lines lie very close together.

The two axes (dimensions) in our 2D spectra are thus both frequency axes. We shall see as we continue that we can adjust our experiment so as to choose different types of frequency information. An early experiment, known as the **J-resolved experiment**, was designed in such a way that one axis was the (proton or carbon) chemical shift axis and the other the one-bond proton-carbon coupling constant. However, this experiment is not generally very useful for structural determination, so that we shall not discuss it here.

The important experiments for our purposes are the *correlation experiments*, where both axes are chemical shift axes. Certainly the most useful of these is the proton-proton correlation experiment, initially known as **COSY** (for **CO**rrelated Spectroscop**Y**) and now, to make things more precise, as **H,H COSY**. This experiment is important, as it provides direct information on which proton nuclei couple with which.

Of course other correlations can be carried out involving any two NMR-active nuclei. The result of a **P,H correlation experiment** will be discussed below. But since most organic molecules do not contain NMR-active nuclei apart from the proton and carbon-13 (or if they do, then certainly not in 100% abundance, with the exception of fluorine-19), the other most important correlation experiments involve C and H as the relevant nuclei. These experiments, the **C-H correlation** (which can be carried out in different ways, although we shall not go into these) tell us directly which proton signal corresponds to which carbon signal. As we shall see, this type of experiment can be adjusted according to the value of the C-H coupling constant involved. We can either detect via the one-bond coupling constants or via the much smaller long-range coupling constants, and we shall see that the results are rather different.

There is also the rather famous experiment known as **2D INADEQUATE** (Incredible Natural Abundance DoublE QUAntum Transfer Experiment) which allows us to correlate carbon-13 with carbon-13. Potentially this experiment is very useful, since it allows us to see directly which carbon atoms are directly bonded. However, you will remember that the natural abundance of carbon-13 is only 1.1%, so a carbon-13/carbon-13 correlation requires us to detect only about 0.01% of the carbon nuclei present. Thus the experiment is very insensitive and requires large amounts of both sample and measuring time (up to 24 hours!). Since phosphorus-31 has a natural abundance of 100%, a **P,C cor-**

relation experiment can be carried out much more quickly, and an example is shown below.

When we think a little more about what happens during a 2D experiment, we realise that it involves the collection and Fourier transformation of a huge amount of data. When 2D experiments were first devised, they were by no means routine. In those days computers were much slower and had much less memory. So the generation of a 2D spectrum involved several hours of measurement and quite a lot of computer time to calculate it from the raw data. Nowadays computers are much faster and have much more memory, so that 2D spectra such as H,H COSY and C-H correlation have become routine. Although we do not want to go into detail about NMR theory, we should mention that two advances in instrumentation have made 2D really fast. One is **inverse detection** (here the carbon-13 information is transferred via carbon-hydrogen coupling to the protons and the much more sensitive proton signal detected) and the other is **gradient spectroscopy** (normally we need to keep the magnetic field across the sample homogeneous, but in certain cases the application of inhomogeneous "gradient pulses", as used in medicinal NMR applications, make NMR experiments much faster). A combination of the two techniques, which is fast becoming state-of-the-art, allows us to carry out the two invaluable H,H COSY and C-H correlation experiments in minutes rather than hours! This is why we shall include them in the majority of the problems in Part 2.

In principle it is possible with many modern spectrometers to carry out correlation experiments using any two NMR-active nuclei, and we shall demonstrate this below by discussing P,C and P,P correlations.

2.2
H,H COSY

The H,H COSY spectrum of model compound **1** is shown in Fig. 24. In fact you can see a total of three spectra: the "central square" which is the actual 2D spectrum and two proton 1D spectra at the top and on the left. The computer software generates this combination of spectra automatically using a previously recorded 1D proton spectrum.

A glance at the proton spectra shows that the OH proton is missing, and when we look at the numbers along the axes we can see that in fact only the range from about 1.2 to 7.2 ppm is covered. This is a principle of 2D: only record the part of the spectrum which contains useful information! Since we want to find out which nucleus couples with which, we do not need to record the OH signal as we already know that it is a singlet.

When we discussed the 1D proton spectrum of model compound **1**, we used decoupling techniques to interpret the coupling patterns. The 2D spectrum allows us to re-check our earlier conclusion. We see a singlet and three multiplets in both proton spectra. If we now look at the central square, we observe a set of signals along a diagonal which we can draw from bottom left to top right.

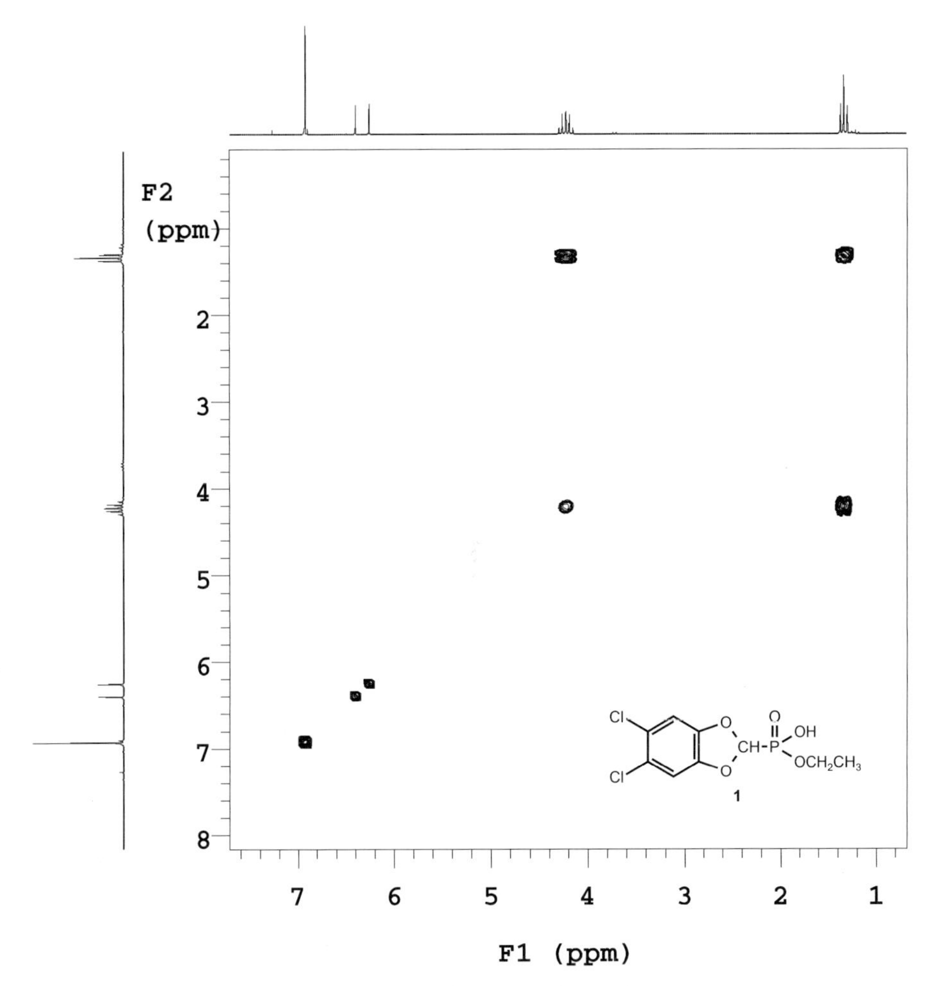

Fig. 24 H,H COSY spectrum (200 MHz) of compound **1** in $CDCl_3$, measurement time 3.5 minutes

When interpreting an H,H COSY spectrum, the first step is to draw in this diagonal and identify the signals lying on it (if we are unlucky, one or more might be missing, and then we would have to adjust the height of the "contour line" using the computer).

Here all the four signals, the singlet and three multiplets, are present on the diagonal. Now we need to locate the information *on the coupling between the protons*. The doublet is caused by the proton-phosphorus coupling, you will remember, so this coupling should not be "active" in the H,H correlation spectrum. But the methyl and the methylene multiplets involve H,H coupling, which should make itself visible. It does so in the form of the **"cross peaks"** or off-diagonal peaks, and we can see two of these, one above and one below the diagonal. These are completely symmetrical with respect to the diagonal, forming a square when we draw lines to connect the signals involved. And this is the secret of interpreting H,H COSY spectra: first draw the diagonal and then locate the squares connecting the peaks on the diagonal with the off-diagonal peaks. As noted above, it is sometimes necessary to shift the contour level to make sure no squares are missing.

2.3 2D NOE

The previous experiment (COSY) demonstrated the interactions (J coupling) between protons via the bonding electrons. The NOE effect which we described in Section 1.1.6 functions because of the through-space interactions between protons, and we used the NOE difference and selective NOE experiments to demonstrate it.

NOE effects can naturally also be investigated by 2D experiments; these are known as **NOESY** and **ROESY**.

We shall use compound **3** to demonstrate the results obtained from a 2D NOESY experiment, and for comparison we shall use the COSY spectrum obtained from the same compound. Fig. 25 shows the COSY (top) and 2D NOESY (bottom) spectra of compound **3**.

In the COSY spectrum we can see a diagonal and two sets of cross peaks: at high field those due to the coupling between CH_2 and CH_3, and at low field those due to the couplings within the aromatic ring.

Now let us look at the NOESY spectrum: just as in COSY, we can identify a diagonal and a series of associated off-diagonal cross peaks. Thus the interpretation of the results is analogous to the method we have already learned for COSY. However, the cross peaks are not due to spin-spin coupling but to NOE effects between the protons concerned. However, if we look more closely we can see one big difference between the diagonal peaks, which look like irregular circles, and the cross peaks, which look just like all the peaks in the COSY spectrum.

The reason for this is that our experiments are **phase-sensitive**. What do we mean by this? You will remember that in the DEPT and APT spectra the CH/CH_3 and CH_2 peaks are in one case positive (up) and in the other negative (down), which we also refer to as in opposite phase. Here in COSY and NOESY our experiments include such phase information, which is read off from the way the signals look in the plot.

Thus all the signals in the COSY spectrum are of the same phase, while in NOESY the diagonal and cross peaks have opposite phase.

We can see three sets of cross peaks: methyl/methylene and aromatic CH as in COSY, and in addition a clear interaction between the methine proton and the aromatic protons closest to it. This interaction is naturally not visible in the COSY spectrum, as the protons are separated by five bonds. A look back to Sec-

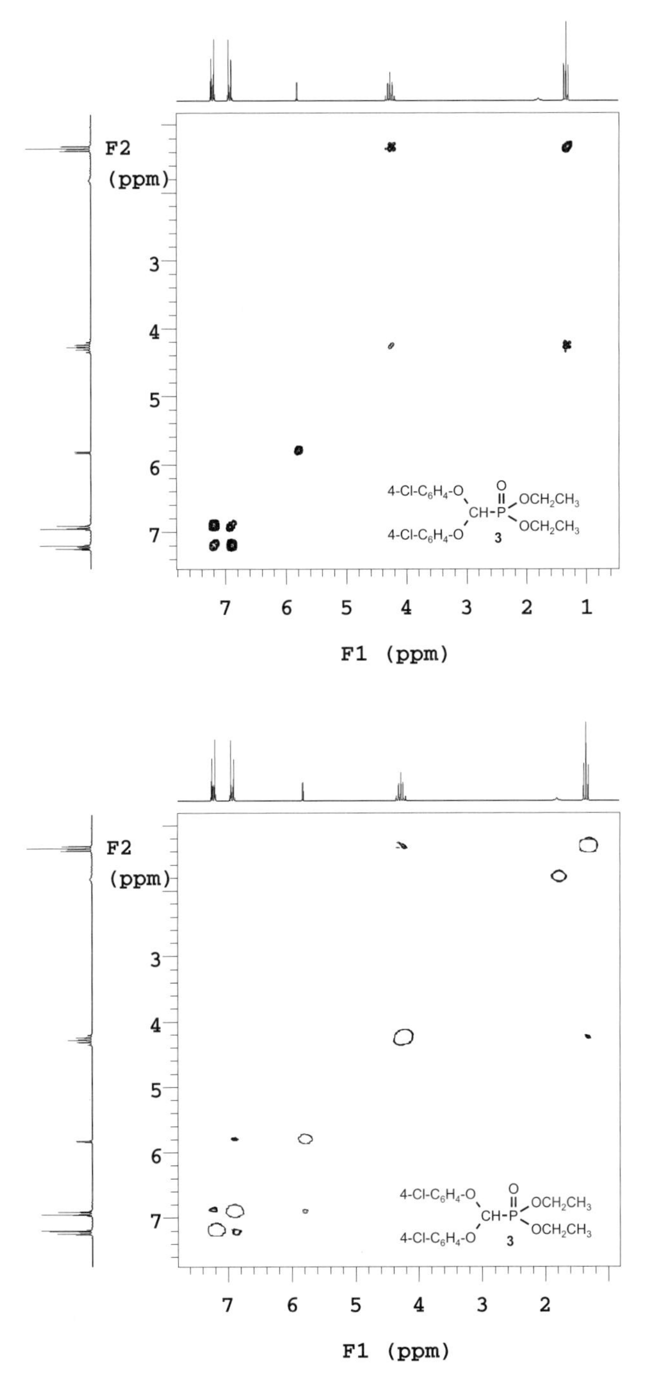

Fig. 25 2D spectra of compound **3**. Top: COSY (200 MHz, $CDCl_3$, measurement time 15 minutes); below: NOESY spectrum (200 MHz, $CDCl_3$, measurement time 40 minutes)

tion 1.1.6 shows that this NOE was (as must be the case) also visible in the 1D experiment.

The advantage of the 2D NOE experiment over selective 1D NOE measurements is that all the NOEs present in a compound are detected in one single experiment, although this takes a relatively long time. Disadvantages of 2D NOE experiments lie in the occurrence of artefacts and problems with the phase correction.

However, 2D NOE studies are invaluable in structure determination, in particular of peptides and proteins: here the NOEs give invaluable information for conformational analysis and the determination of the tertiary structures of proteins.

2.4
P,H COSY: With Varying Mixing Times for the Coupling

Since phosphorus and protons are both abundant spin-1/2 nuclei, it is simple to design an experiment in which we correlate protons and phosphorus rather than protons with themselves. The result of this experiment, a P,H correlation,

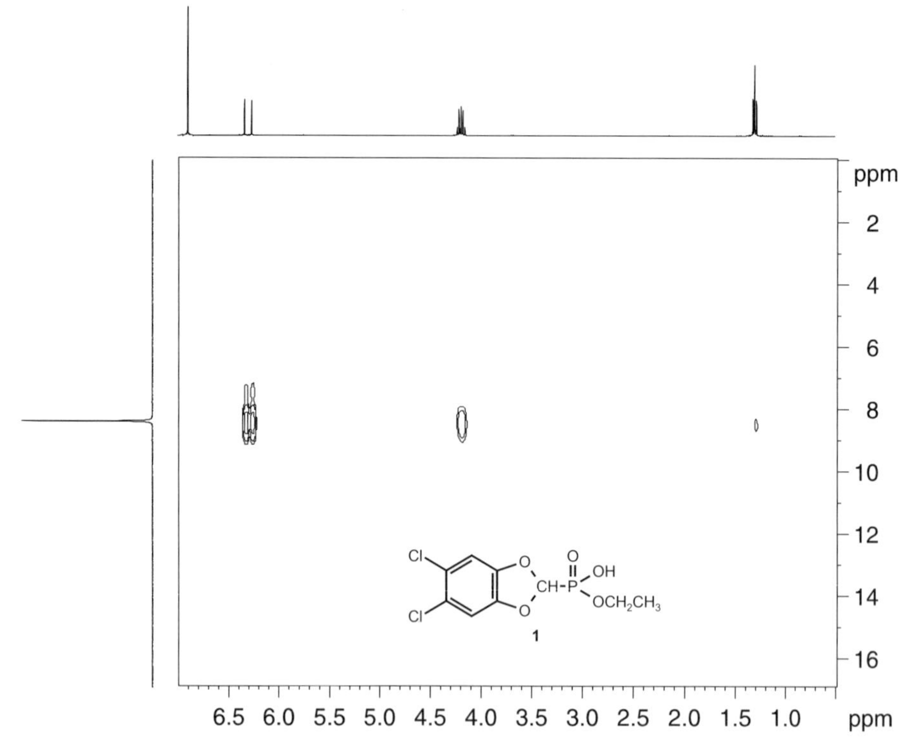

Fig. 26 P,H correlation spectrum of compound **1** (400 MHz, 5% in $CDCl_3$, delay set for J_{PH} = 1.65 Hz, measurement time 12 minutes)

is shown in Fig. 26. Again we have the 2D spectrum in the form of a central rectangle and two (previously recorded) 1D spectra parallel to the axes. One is the proton spectrum, the other the phosphorus spectrum. The latter of course consists of a single line, and in the 2D spectrum we do not need to look for a diagonal as there cannot be one.

Instead, there are three rather broad contour signals corresponding to the coupling of the phosphorus with (from left to right) the methine, methylene and methyl protons. The breadth of the signals is roughly proportional to the magnitude of the coupling J involved.

This is not surprising, as the input of an average coupling constant is part of the set-up of the experiment. In fact the time period 1/J (in seconds) is involved in the experiment, and 1/J increases as J decreases. If the individual experiment is too long, the signal intensity will be decreased by relaxation.

2.5 C,H Direct Correlation

Organic compounds almost always contain carbon and hydrogen, so that the C,H correlation is a key experiment in every structural determination. This experiment tells us which carbon signal corresponds to which proton signal, and the result for model compound **1** is shown in Fig. 27.

By now we are used to the appearance of such spectra, and again the central rectangle contains the actual 2D spectrum, while the carbon spectrum (decoupled) is shown on the left and the proton spectrum at the top.

Do not try to draw a diagonal; there is none. Probably the best thing to do when you are dealing with an unknown molecule is to construct a table, which in the present case could look like this:

H signal at	correlates with	C signal at
6.9 (aromatic)		110
6.3 (methine doublet)		106
4.2 (methylene multiplet)		65
1.3 (methyl triplet)		17

Note that, apart from the solvent signal, two aromatic carbon signals (at 125 and 147 ppm) show no correlation because they are quaternary (i.e. not bonded to protons).

Again, we need to define a coupling constant J to set up this experiment. Here for optimum sensitivity we have used an average value for direct (one-bond) carbon-hydrogen coupling constants of 140 Hz. This choice works well for most CH bonds, but is rather low if an acetylenic CH bond is present.

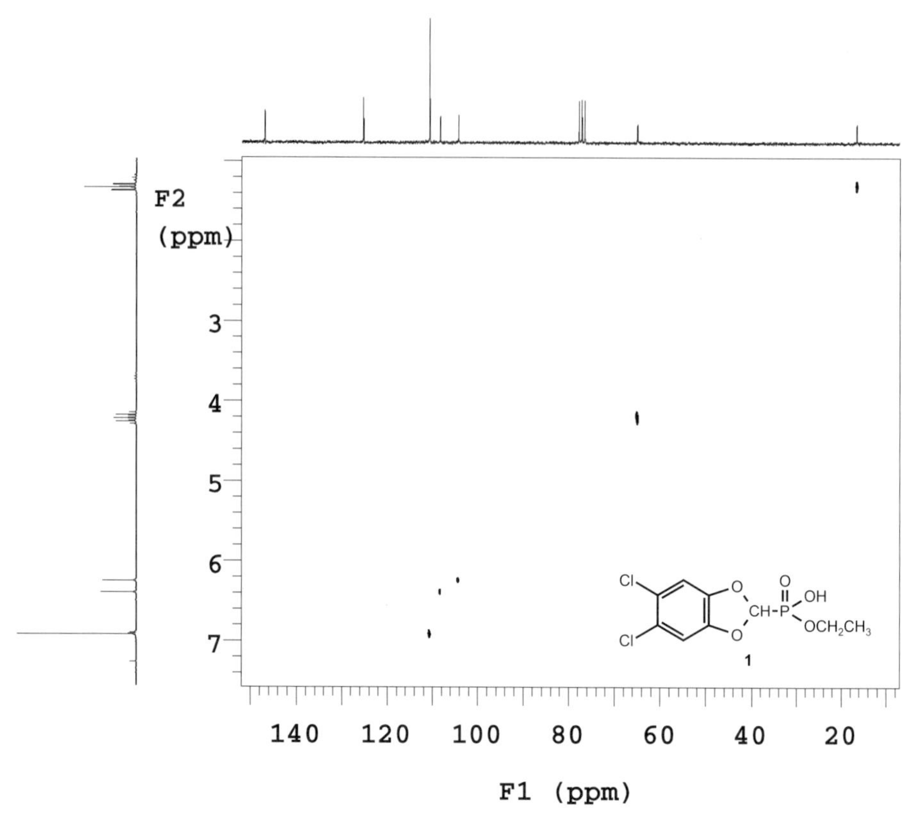

Fig. 27 C,H correlation spectrum for compound **1** (set for directly bonded hydrogens, 200 MHz, 5% in $CDCl_3$, measurement time 60 minutes, inverse detection)

2.6 C,H Long Range Correlation

We can vary the result of a C,H correlation experiment by varying the coupling constant value we use (this is called varying the **"mixing time"**). Carbon also couples to hydrogen across two or three bonds (sometimes more), but the coupling constants are drastically smaller than the one-bond coupling constant. The spectrum of **1** shown in Fig. 28 results from a long-range experiment; here J has been set to 8 Hz, which means that each single experiment is longer than in the direct experiment and that due to relaxation the signal accumulated in each experiment is smaller. However, the combination of inverse detection and gradient application makes the experiment fast. This technique is not used routinely but only to answer specific questions about the molecule under study. Let us construct a table for this experiment:

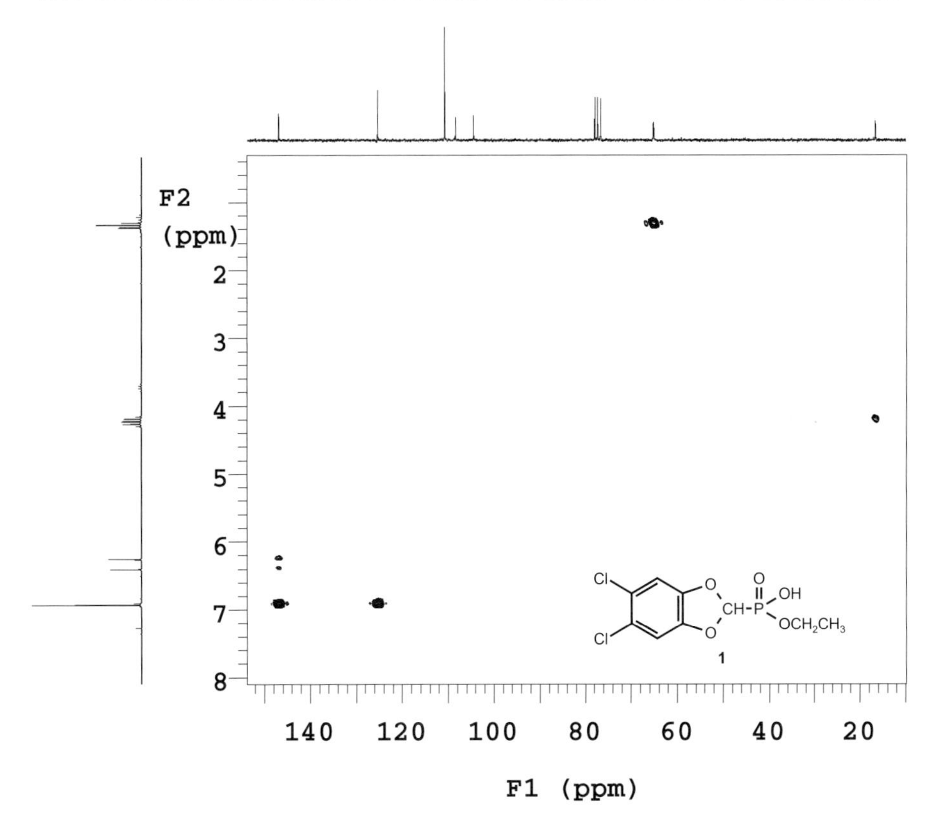

Fig. 28 C,H correlation spectrum for compound **1** (set for long-range coupling, 200 MHz, 5% in $CDCl_3$, measurement time 60 minutes, inverse detection)

H signal at	correlates with	C signals at
6.9 (aromatic)		147, 125
6.3 (methine doublet)		147
4.2 (methylene multiplet)		17
1.3 (methyl triplet)		65

Because we are detecting via long-range coupling, the correlations to the methyl and methylene signals are reversed. The aromatic CH signal in the proton spectrum now correlates with both quaternary carbons, as expected. The methine doublet in the proton spectrum, however, correlates weakly with *only one* of the two low-field quaternary aromatic carbon signals; we can thus make a clear assignment of these carbons.

To be fair, we must point out that this type of experiment is extremely sensitive to the parameters chosen. Various pulse sequences are available, including the original ***COLOC*** *(COrrelation by means of LOng range Coupling) as well as* *experiments variously referred to as* ***HMBC*** *(Heteronuclear Multiple-Bond Correlation) and* ***HMQC*** *(Heteronuclear Multiple-Quantum Correlation). Depending on the parameters chosen, it is often not possible to suppress correlations due to one-bond coupling!*

2.7 P,C Correlation

A P,C correlation experiment also requires that we use a predefined coupling constant value to determine the mixing time. A brief look at the proton decoupled carbon-13 spectrum (Fig.14) shows that $^{1}J_{PC}$ is very large (around 200 Hz), while the long-range J_{PC} values are much smaller (around 5–10 Hz).

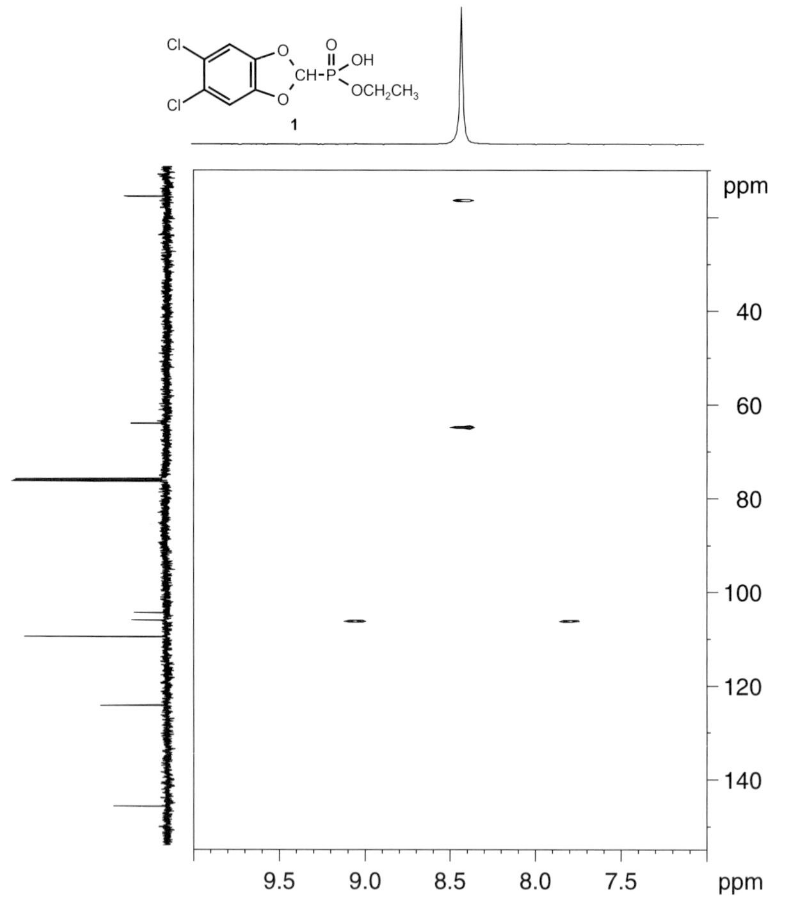

Fig. 29 P,C correlation spectrum for compound **1** (set for long-range coupling, 400 MHz, 5% in $CDCl_3$, inverse detection)

Figure 29 shows the P,C correlation for compound **1** carried out by selecting a J value of 15 Hz. The decoupled phosphorus signal is shown at the top, the proton decoupled carbon-13 spectrum on the left. The actual 2D spectrum is in the centre and is as we might expect very simple (naturally there is no diagonal).

The first thing that we can see is that the 2D spectrum is *not* decoupled with respect to the phosphorus: the methine carbon C-P doublet in the ^{13}C spectrum is associated with a doublet along the phosphorus axis, from which $^{1}J_{PC}$ can of course be extracted.

Two other carbons show correlations, the methylene and methyl signals.

No correlations to aromatic carbons are visible, although the ^{13}C spectrum in Fig. 14 shows that the aromatic carbons bonded to oxygen do couple with phosphorus: if we were to carry out a second experiment using a smaller J value this correlation would however become visible.

2.8 P,P Correlation

We mentioned above that it is possible to carry out carbon-carbon correlation experiments using the 2D **INADEQUATE** procedure. There, as you may remember from the discussion of one-dimensional INADEQUATE, we have a very difficult problem to solve: carbon-13 represents only 1.1% of the total carbon nuclei present in a sample. And in the INADEQUATE experiment we need to detect only those molecules containing two carbon nuclei which couple with one another, i.e. about 10^{-4} of the nuclei present; at the same time we have to get rid of the signal coming from those molecules containing only one carbon-13 (the great majority)!

As ^{31}P is the only phosphorus isotope, these problems do not arise. Figure 30 shows the result of a 2D P,P correlation carried out for compound **6**.

Now, as in the H,H correlation, we have a central square and a diagonal, and have to look for a squares arising from the diagonal and cross peaks.

We immediately see the squares arising from the couplings between ^{c}P and ^{b}P and between ^{b}P and ^{a}P. But if we look closely we can see that there is also a weak correlation between ^{c}P and ^{a}P: this shows that there *is* a coupling between them, as we had expected. But because we could not see the coupling in the 1D spectrum the coupling constant must be smaller than the signal linewidth. This is one of the beauties of 2D correlation experiments: they often allow the detection of couplings which are not visible in the corresponding 1D spectra!

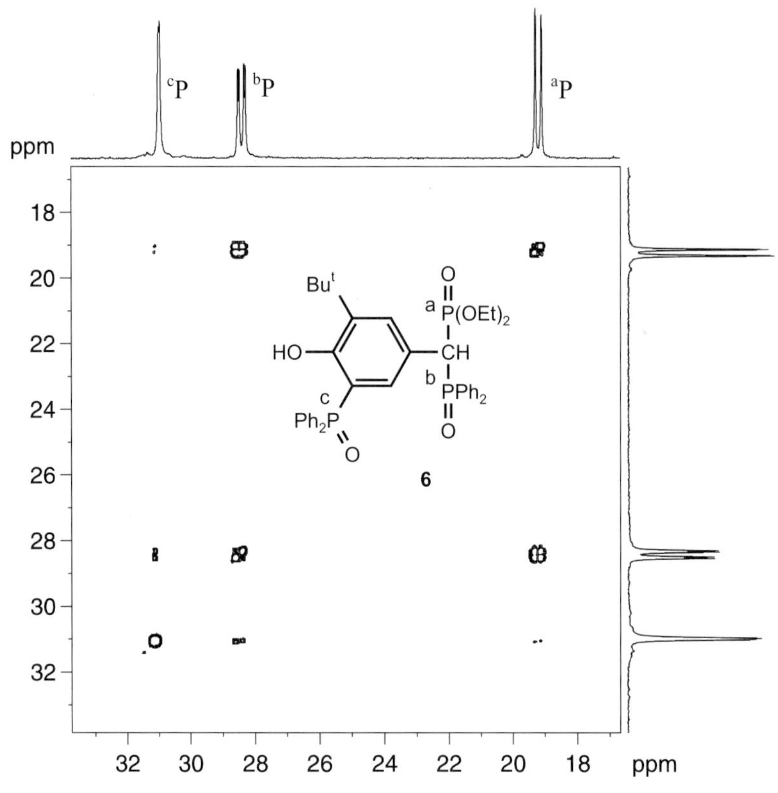

Fig. 30 P,P correlation spectrum for compound **6** (202 MHz, measurement time 15 minutes)

3 Quadrupolar Nucleus Experiments

3.1 General Principles: Quadrupole Moment, Relaxation, Linewidth

The experiments we have so far described have been used to study nuclei with spin I = ½ (^{1}H, ^{13}C, ^{31}P). Our model compounds **1** and **2** contain two further atoms (oxygen and chlorine), which have no NMR-active isotope with spin ½. Oxygen does however have an NMR-active isotope with spin I = 5/2 but very low natural abundance (0.037%): this is ^{17}O. Chlorine has two NMR-active isotopes: ^{35}Cl (I = 3/2, 75.53%) and ^{37}Cl (I = 3/2, 24.47%).

NMR-active nuclei with spin > 1/2 (these include, as we mentioned previously, deuterium) have an electric quadrupole moment and are thus referred to as **quadrupolar nuclei.**

These nuclei (and they form by far the majority of the NMR-active nuclei!) are subject to relaxation mechanisms which involve interactions with the quadrupole moment. The relaxation times T_1 and T_2 of such nuclei are very short, so that very broad NMR lines are normally observed. The relaxation times, and the linewidths, depend on the symmetry of the electronic environment. If the charge distribution is spherically symmetrical the lines are sharp, but if it is ellipsoidal they are broad.

3.2 ^{17}O

Oxygen plays a central role in organic and inorganic chemistry as well as a vital role in animal and plant life. NMR studies on this element could therefore be of great interest. Although oxygen-17 has such a low natural abundance, it is possible under correctly chosen conditions to obtain high-quality NMR spectra. Thus NMR measurements on biological materials can readily be carried out. The chemical shift range is very large (around 2500 ppm), so that in spite of the large linewidths observed it is possible to study structural changes readily: coupling information can also often be obtained.

Briefly, the experimental conditions should be based on the following information: acetonitrile is the recommended solvent, as it gives sharper lines than chloroform. Temperature also affects the linewidth, so that the effect of increasing it above room temperature should be tested. Because of the fast relaxation of the oxygen nuclei it is possible to use extremely short pulse repetition rates (50–200 ms), and the acquisition time should also be made short by appropriate choice of the number of data points and the sweep width. In this way we can record a large number of FIDs in a relatively short time. The FID needs to be subjected to exponential multiplication using linewidths of 50 to 500 Hz.

The normal reference substance is water, the signal of which is set equal to 0 ppm.

3.2.1 *^{17}O Spectrum of 7: Chemical Shift (Reference), Coupling with P*

We shall use our model compound **7** to show how oxygen-17 NMR can be used. Figure 31 shows the spectrum, recorded using a 40% solution of **4** in CD_3CN at a temperature of 55°C.

Compound 7 contains four different oxygen nuclei, so that we expect four signals. As in carbon-13 NMR, the signal associated with the carbonyl group lies at very low field (364 ppm). The signal for the P=O oxygen at 99 ppm is immediately recognisable because of the presence of the one-bond P-O coupling (J = 174 Hz). The two remaining signals are due to the methoxy oxygen bound to carbon and the ethoxy oxygens bound to phosphorus: here the signal intensity differ-

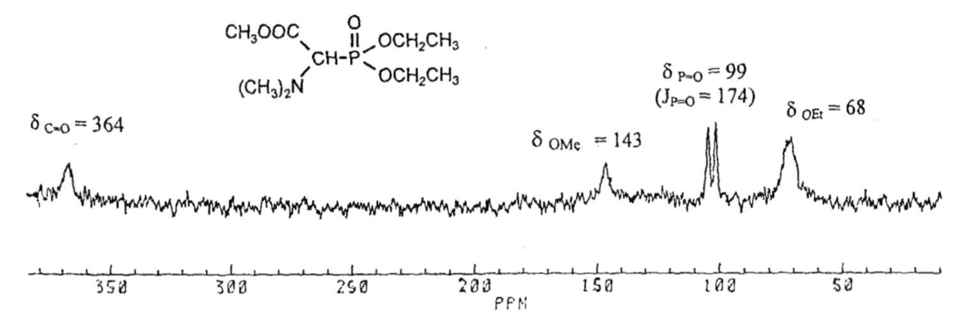

Fig. 31 Oxygen-17 spectrum for compound 7 (40% in CD_3CN, temperature 55°C)

ence indicates which is which, and the literature confirms that the high-field signal is indeed due to the ethoxy oxygens.

3.2.2
P-O Correlation

In principle it is possible (with a suitably configured spectrometer) to carry out correlation experiments between any pair of NMR-active nuclei. However, a P-O correlation is certainly not trivial, as we are dealing with a "good" (spin-1/2) and a "bad" (quadrupolar) nucleus. Indeed, all our attempts to carry out such an experiment failed completely, and as far as we are aware no such experiments have so far been reported in the chemical literature.

4
HPLC-NMR Coupling

4.1
General Principles, NMR as a Highly Sensitive Analytical Tool (µg to ng Amounts)

The identification and structural characterisation of biological materials, obtained for example from plants, was traditionally carried out via the classical sequence involving extraction, separation, isolation and characterisation, a sequence which requires large amounts of substance and a great deal of time. Industrial problems, for example the search for small amounts of contaminants in industrial products or in waste water, also require intensive analytical studies.

A direct combination of separation and analysis techniques is thus invaluable. **GLC-MS** and **HPLC-MS** coupling are now routinely used. Because of the high sensitivity of modern NMR instruments the coupling of HPLC and NMR is now used in many NMR laboratories, and we shall discuss the principles and show some results below.

The coupling of HPLC in tandem with NMR requires two separate systems:

a) a conventional HPLC system with a standard detector (e.g. UV) and a monitoring system to observe and control the chromatography.
b) a normal NMR spectrometer with a dedicated probehead.

A long capillary with a computer-controlled switching valve (the instruments must be separated by 2–3 metres because of the strong magnetic field) connects the exit from the HPLC with the probehead. The latter is completely different in its construction from conventional probeheads: instead of the NMR tube there is a small flow cell, the volume of which is 40–100μl. The transmitter and receiver coils are attached directly to the cell in order to maximize the sensitivity.

There are two different ways of carrying out an HPLC-NMR experiment:

Continuous Flow:
The NMR spectrum is recorded during the chromatographic separation. Data are collected as in a 2D experiment, the two dimensions being the chemical shift and the retention time of the chromatogram.

Stopped Flow:
Here the chromatographic scan is stopped at defined times and the NMR experiments then carried out. In this case it is possible to adjust the measurement time of the experiment to the concentration of the sample.

It is normal in conventional NMR to work with deuterated solvents, which serve both for optimising magnetic field homogeneity (lock, shim) and for avoiding the presence of the unwanted strong signals from protonated solvents.

HPLC requires much larger amounts of solvents, so that deuterated materials are too expensive; instead we work with undeuterated HPLC quality solvents, the proton signals from which are suppressed using the so-called WET sequence, which also suppress the carbon-13 satellites of the solvent signals.

Magnetic field homogeneity is ensured by the presence of D_2O at the beginning of the experiment, since many chromatographic separations use water as one solvent component. Once the homogeneity has been optimised, the coupling experiment can be carried out by changing solvent composition by replacing D_2O by pure water.

4.2 Example: Separation of 4 and 5, two Acetals of Formylphosphonic Ester

The reaction between the cyclic orthoester **8** and diethyl chlorophosphite **9** leads via transesterification to the two acetals **4** and **5**, which cannot be separated by distillation.

$$\underset{\mathbf{8}}{(CH_2O)_2CH{-}OCH_2CH_3} + \underset{\mathbf{9}}{(CH_3CH_2O)_2P{-}Cl} \xrightarrow[-\,CH_3CH_2Cl]{} \underset{\mathbf{4}}{(CH_3CH_2O)_2CH{-}P(O)(OCH_2CH_3)_2} + \underset{\mathbf{5}}{(CH_2O)_2CH{-}P(O)(OCH_2CH_3)_2}$$

4.3 Chromatogram

Figure 32 shows the HPLC chromatogram of the reaction mixture using acetonitrile and D_2O (80:20) as solvent.

Three main peaks (1, 8 and 11) can be seen, two of which are the acetals and the third an unknown by-product.

The next step is to set the same conditions for the HPLC system which is coupled with the NMR spectrometer. The field homogeneity of the probehead is first optimised (shimmed) using the same separation column and solvent mixture.

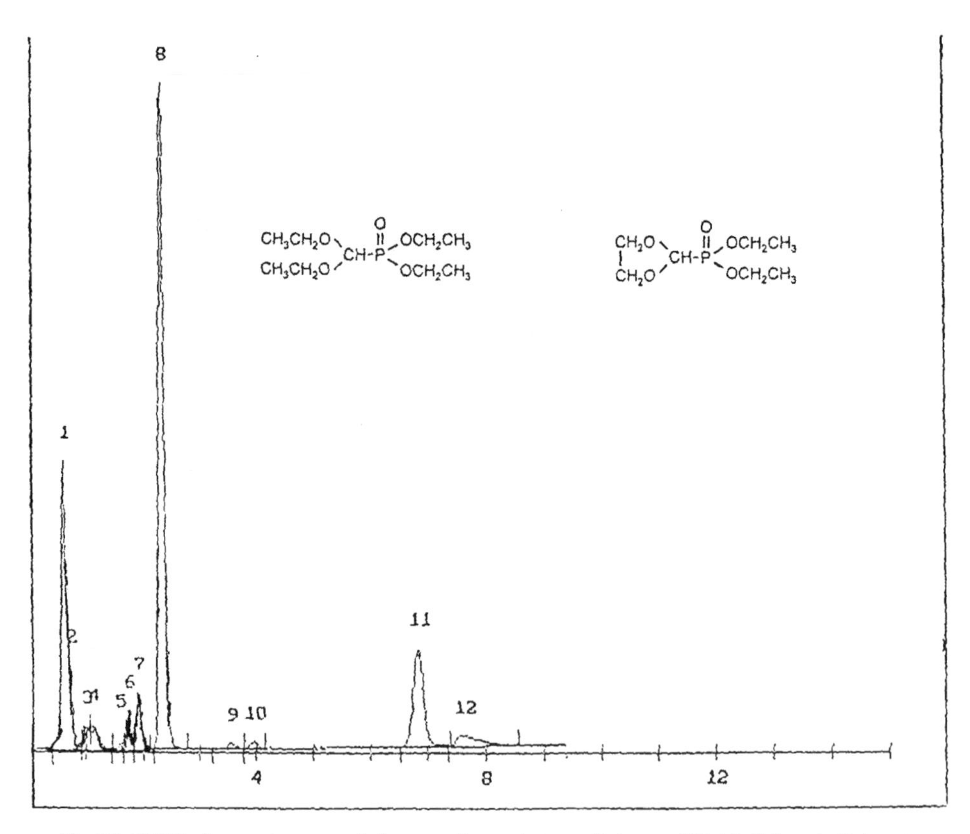

Fig. 32 HPLC chromatogram of the reaction mixture. Column: RP 18 (15 cm). Solvent: CH_3CN/D_2O (80:20). UV detector 190 nm. Pressure 157 bar. Flow rate: 1 mL/min

4.4
On-flow Diagram (Chemical Shift vs. Time)

An on-flow experiment is now carried out. 50 µl of a solution of the product mixture (5 mg in 5 mL solvent) are injected and the NMR proton signal accumulation started simultaneously. The time taken for the chromatogram is 17 min. During this time a total of 128 proton NMR spectra are recorded, each with 8 scans, i.e. an FID is accumulated approximately every 7 s. After the Fourier transformation we obtain a two-dimensional representation (Fig. 33) of the on-flow experiment.

The two axes represent the chromatogram (retention time) and the chemical shift information. The individual NMR spectra can be extracted by the software and viewed individually in the form of normal 1D proton NMR spectra.

The spectra recorded at retention times 1, 3 and 7.5 minutes are shown in Fig. 34.

The lower spectrum shows an ester group (triplet at 1.2 ppm, quartet at 4.2 ppm) and a singlet at 8.1 ppm, the latter indicating the presence of a formyl group. Peak 1 results from ethyl formate, formed by adventitious hydrolysis of the acetal **4.** The middle and top spectra correspond to the two acetals (see equation); the assignment is very simple.

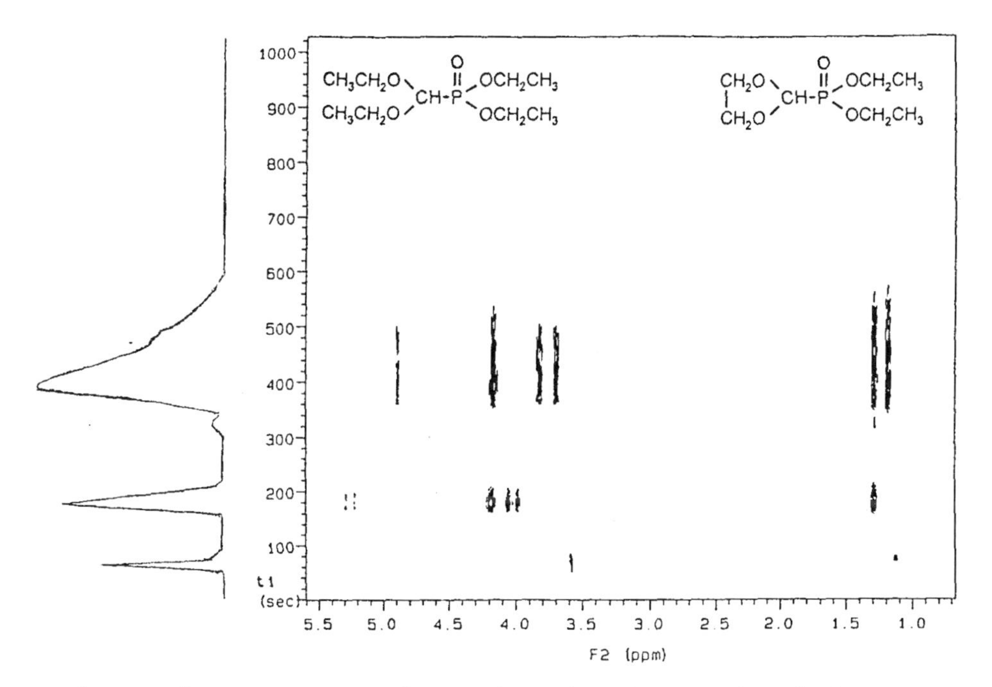

Fig. 33 On-flow experiment carried out on the product mixture. Horizontal axis: proton chemical shift. Vertical axis: retention time

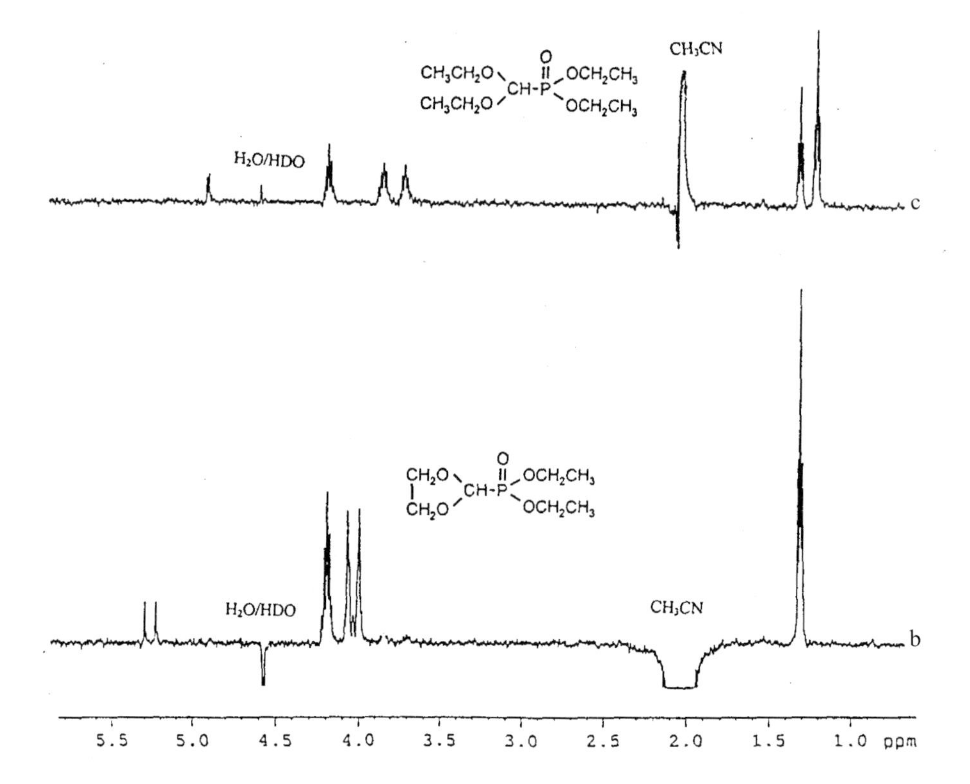

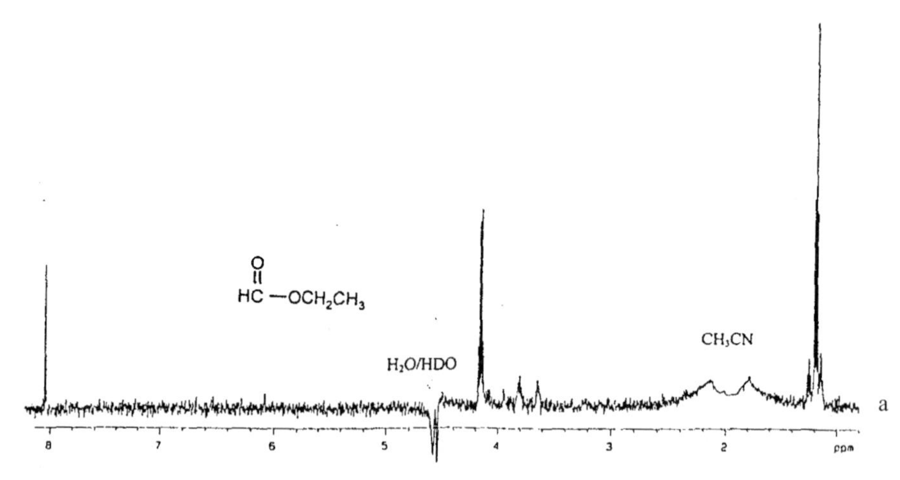

Fig. 34 Proton NMR spectra from the on-flow experiment. FIDs recorded after **a** 1 min., **b** 3 min., **c** 7.5 min.

Acetal **5** contains only one type of methyl group, while acetal **4** has two different types. Thus the spectrum recorded after 3 minutes corresponds to acetal **5**, while that recorded after 7.5 minutes results from acetal **4**. The methylene groups (between 3.6 and 4.2 ppm) and the methine protons (at 4.8 and 5.4 respectively) confirm these assignments.

4.5 Stopped Flow Experiments

Spectra which are better resolved (useful for example for the exact determination of coupling constants) can be obtained by carrying out stopped-flow experiments. Here we stop the chromatographic separation after 3 and 7.5 minutes, optimise the homogeneity (by shimming the magnet) and carry out the desired NMR experiments.

Figure 35 shows the proton spectra which we obtain; you can see that they are of much better quality than those we got from the on-flow experiment. The signals for acetonitrile and residual HDO have been cleanly removed using the WET sequence referred to above, and resolution and signal-noise are much better, so we can obtain coupling constants exactly.

The stopped-flow technique also allows us to obtain spectra which require relatively long measuring times: as an example we show the H,H-COSY spectrum (Fig. 36) of peak 8 (retention time 3 minutes).

The experiment took 11 minutes, and the spectrum shows quite clearly the correlation signals for the acetal **5**: cross peaks between the methyl and methylene signals from the ethyl group and between the magnetically non-equivalent protons of the ethylene bridge. CH correlation experiments can easily also

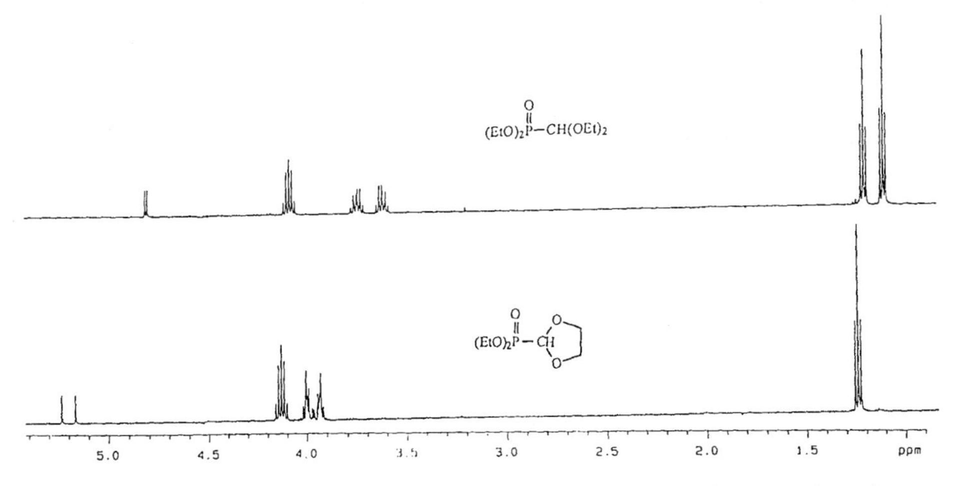

Fig. 35 Proton spectra obtained from the stopped-flow experiment. Above: acetal **4**. Below: acetal **5**. In each case 16 scans, relaxation delay 1 s

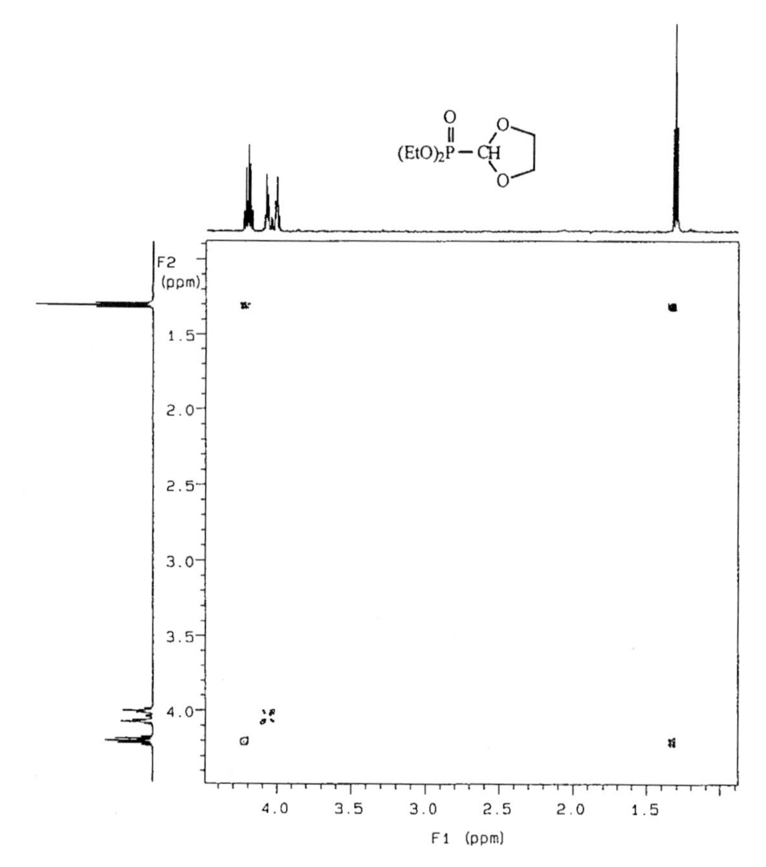

Fig. 36 COSY spectrum of acetal **5** obtained in a stopped-flow experiment. Measurement time 11 min

be carried out, even though in the case of the two acetals **4** and **5** they require between two and three hours!

Appendix: Reference List

Basic but thorough

NMR Spectroscopy. 2nd Ed.; H. Günther, John Wiley and Sons, Chichester, 1995. ISBN 0-471-95199-4 (cloth)/0-471-95201-X (paper).

Basic One- and Two-Dimensional NMR Spectroscopy. 2nd Edition; H. Friebolin, VCH, New York, 1998. ISBN 3-527-29513-5.

More advanced

Modern NMR techniques for Chemistry Research; A.E. Derome, Pergamon Press, Oxford, 1987. ISBN 0-08-032514-9 (Hardcover)/0-08-032513-0 (Softcover).

Modern NMR Spectroscopy, 2nd. Edition; J.K.M. Sanders and B.K. Hunter, Oxford University Press, Oxford, 1993. ISBN 0-19-855566-0.

Invaluable for solving structural problems

Structure Determination of Organic Compounds. Tables of Spectral Data. 3rd Ed.; E. Pretsch, P. Bühlmann and C. Affolter, Springer, Berlin 2002. ISBN 3-540-67815-8 (Softcover, with CD-ROM).

Spektroskopische Daten zur Strukturaufklärung organischer Verbindungen. 4. Aufl.; E. Pretsch, P. Bühlmann, C. Affolter, M. Badertscher, Springer, Berlin 2002. ISBN 3-540-41877-6 (Softcover, with CD-Rom).

Part 2: Worked Example and Problems

1
Introduction

Readers can obtain a list of answers to the problems on application (by mail or e-mail) to the authors.

Terence N. Mitchell
Universität Dortmund
– Fachbereich Chemie –
44221 Dortmund
Germany
e-mail: mitchell@chemie.uni-dortmund.de

Burkhard Costisella
Universität Dortmund
– Fachbereich Chemie –
44221 Dortmund
Germany
e-mail: Costi@chemie.uni-dortmund.de

Before starting on the problems, please refer to the worked example (see p. 64); the detailed data given in the headers of each problem should be read through carefully as they will provide vital information!

The book contains 35 problems, ordered according to the complexity of the molecules involved. For each problem the following NMR spectra will generally be reproduced:

Proton spectrum with integration
Carbon-13 spectrum with DEPT or APT for multiplicity information
H,H-COSY
H,C-COSY

For some of the very simplest molecules the 2D spectra are not included.

Additional information is provided in all cases in the form of
a) the molecular formula
b) IR frequencies corresponding to functional groups present (e.g. OH, NH, C=O, NO_2)

Solving the Structures of Organic Molecules

NMR spectroscopy is only one of a series of tools and methods which can be used in order to determine the structures of unknown organic molecules, but since it is by far the most powerful we have decided to concentrate our attention to it in this book. However, in the real world one will not start by running NMR spectra. Preliminary information can be obtained from the following:

Elemental Analysis

If the sample is pure (this can generally be checked by thin layer chromatography or gas chromatography) then the elemental analysis values for carbon, hydrogen and nitrogen can be used to obtain element ratios, provided that C, H, N and O are the only elements present.

Mass Spectrometry

If you are lucky, the ion with the highest mass to charge value will be the molecular ion. However, this is often not the case, as textbooks on mass spectrometry make clear. If it is possible to carry out high resolution mass spectrometry on the molecules in question, and the molecular ion is indeed observed, the exact mass can be used in combination with tables to obtain the molecular formula directly. Alternatively, you can use the internet (http://www.sisweb.com/cgi-bin/mass10.pl) to calculate and plot mass distributions for any molecular fragment you think is present.

In this book, in order that you can concentrate your attention on the NMR spectra, we shall provide you with the molecular formula in all cases. This in turn provides you with information which can be extremely useful during the process of solving the structure: if the molecule only contains C, H, N and O then you can use the molecular formula to obtain the number of so-called double bond equivalents, i.e. information on the degree of unsaturation. Though there are various formulas which can be devised to do this, we recommend the calculation using the following formula: for a molecule $C_aH_bO_cN_d$, the number of double bond equivalents DBE is calculated as follows

$$DBE = [(2a + 2) - (b - d)]/2$$

Oxygen and any other divalent elements present are ignored. Any other monovalent elements present, such as halogens, are treated as hydrogens, any other tetravalent elements (e.g. Si) as carbons. If other trivalent elements are contained in the molecules (this is rather unlikely for most organic molecules, but trivalent phosphorus is one example) they are treated as nitrogens.

We need to define what is meant by a double bond equivalent: any element-element double bond (C=C, C=O, C=N) counts as 1, while triple bonds count as 2. A saturated ring counts as 1, and any double bond present in the ring also counts as 1: thus a benzene ring corresponds to 4 double bond equivalents.

We highly recommend that when solving the problems you make use of a book which contains tables of NMR chemical shifts and coupling constants as well as infrared frequencies. While various suitable texts are available, our preference is the book by Pretsch et al. (see Appendix for details). Though earlier editions are available, we recommend the new 3rd edition, which includes a CD-ROM with NMR prediction software.

These problems have been chosen so that they can be solved using standardised sets of NMR spectra, together with the additional information listed above.

This does not mean that the other techniques discussed in Part 1 are unimportant, but that they will only need to be made use of when the "standard" spectra do not provide sufficient information.

What do we consider to be the standard set of spectra which you should always try to obtain?

The proton spectrum, with integration. This spectrum tells you how many magnetically non-equivalent types of proton are present in the molecule, where they absorb (i.e. which type of proton the signals represent), and the *relative* numbers of protons. Since we shall give you the total number of protons present, you will be able to calculate the *absolute* numbers of protons of various types in the molecule.

The carbon spectrum, both in the broadband decoupled form and as an APT spectrum.
APT, you will perhaps remember, stands for Attached Proton Test, meaning that this spectrum tells you the multiplicity of the signals (Me, CH_2, CH or quaternary C). These two spectra tell you how many magnetically non-equivalent types of carbon are present in the molecule, but (for the reasons we discussed earlier) we do not use integration to try to find out relative numbers. We shall present APT spectra as follows: CH, CH_3 in positive phase (up), CH_2 and quaternary C in negative phase (down).

You may be told that your NMR laboratory does not routinely use APT spectra but provides **DEPT** spectra **(Distortionless Enhancement by Polarisation Transfer)** instead. This is no problem, as DEPT spectra also provide you with the information you need: just go back and read what we have said about the relative merits of APT and DEPT.

The proton-proton COSY spectrum (COSY meaning Correlated SpectroscopY), which tells you directly which protons couple with which. In many cases this information is already available from the proton spectrum, but since multiplets in proton spectra can be quite complicated, even at 400 MHz, the COSY spectra should be recorded as they are very simple to interpret.

The proton-carbon COSY spectrum, which tells you directly which signals in the proton spectrum correspond with which signals in the broadband decoupled carbon spectrum. This information, together with the integration values and the multiplicities obtained from APT (or DEPT), is invaluable in putting together the molecular fragments.

These four NMR spectra will form the basis which you can use to solve the structures (in some cases not all are presented, depending on what information they give). We have naturally arranged the problems on the basis of their molecular complexity, but even very small molecules can have complex proton spectra! All the problems can be solved completely, i.e. including the determination of the isomer involved.

We have used only two different solvents, deuterochloroform ($CDCl_3$) and hexadeuterodimethylsulfoxide (DMSO-d_6). The former dissolves a large majority of organic molecules, but DMSO must be used for more polar substances. The disadvantage of DMSO is that it is very hygroscopic, so that even if you try hard to keep it dry you may find small signals due to water in your spectra. Look out for these in the spectra in this book: they normally lie at about 3.3 ppm.

Solvent shifts are as follows:

Deuterochloroform: ^{1}H, residual $CHCl_3$ in $CDCl_3$ 7.24 ppm (singlet), ^{13}C 77.2 ppm (1:1:1 triplet).

Hexadeuterodimethylsulphoxide: ^{1}H, residual uncompletely deuterated DMSO 2.48 ppm (multiplet), ^{13}C 39.5 ppm (multiplet).

In the real world, chemicals are not always pure; in some cases this is also true of the samples we have used in this book. Thus you may find some small signals due to impurities. These can easily be identified because they are not “labelled” with chemical shift and signal intensity (integration) information.

Before we leave you to start on the problems, we feel that it is vital to present a worked example. Though of course you can work as you like, we highly recommend that you try to follow a relatively standard procedure, which will allow you to put together your facts and deductions in a systematic way and thus make it easier for you to arrive at the correct solution in a minimum time.

2 Worked Example

First use the molecular formula and the equation given above to calculate the number of double bond equivalents. In this case (remembering to treat bromine as equivalent to hydrogen) the value is 1. The infrared spectrum shows a band at 1641 cm^{-1}, which probably represents the C=C bond stretch, but in this case there *can* only be a C=C bond present!

This bond is clearly visible in both the proton and carbon spectra. We recommend making a table of the information these give; the tables can be added to as the structure elucidation continues.

Proton NMR:

5.8 ppm	multiplet	1H	olefinic H
5.1 ppm	multiplet	2H	olefinic H
3.4 ppm	triplet	2H	aliphatic H
2.6 ppm	quartet with fine structure	2H	aliphatic H

Carbon-13 NMR:

135.4 ppm	olefinic CH (signal in positive phase)
117.7 ppm	olefinic CH_2
37.2 ppm	aliphatic CH_2
32.2 ppm	aliphatic CH_2

The fact that we have *three* olefinic hydrogens means that our compound is a terminal olefin, the fact that the other two carbons are both methylene carbons means that our substituent, bromine, is terminal. Thus the only possibility we have is that we are dealing with 4-bromo-1-butene (try to find another isomer that fits!). But this simple molecule has a highly complex proton spectrum, which can only be interpreted completely (exact chemical shift, coupling constants) by spectrum simulation.

However, we have already obtained the structure, which is

H_2C=CH-CH_2-CH_2Br

In this case we do not really need the 2D spectra, but we should take the time to look at and interpret them to see how it is done.

H,H Correlation

First draw the diagonal to make interpretation easier. Then label your hydrogens (from left to right: 2, 1, 4, 3). Now look for and (if you like draw) the squares which demonstrate the couplings present. It is immediately obvious that 1 and 2 couple and that 3 and 4 couple (if you did not have the proton assignment, it would be wise to construct another table to show which couplings are present). Since 1 and 2, and also 3 and 4, are separated by three bonds we obviously expect to see coupling.

But we can find two other squares, involving 2 and 3, and 1 and 3. No further square involving 4 is present. When we think about his for a moment, it is obvious (the expert would say "of course 1, 2 and 3 all couple with one another, because it is an allylic system"). If the additional couplings were not present, the number of lines would be much smaller!

C,H Correlation

This time there is no diagonal. You can label the hydrogens again, but how about the carbons? Is it 2, 1, 3, 4 or 2, 1, 4, 3? The beauty of this correlation is that it gives us the answer straight away. The $BrCH_2$ triplet (H-4) corresponds to the highest-field carbon signal, while the H-3 multiplet corresponds to the carbon signal at 37.2 ppm. Thus the order is 2, 1, 3, 4.

If you try to work in the way we have indicated, you should be able to solve all the problems in this book. But do not try to work through them in order: if you feel lost with one of them, just try the next! Good luck!

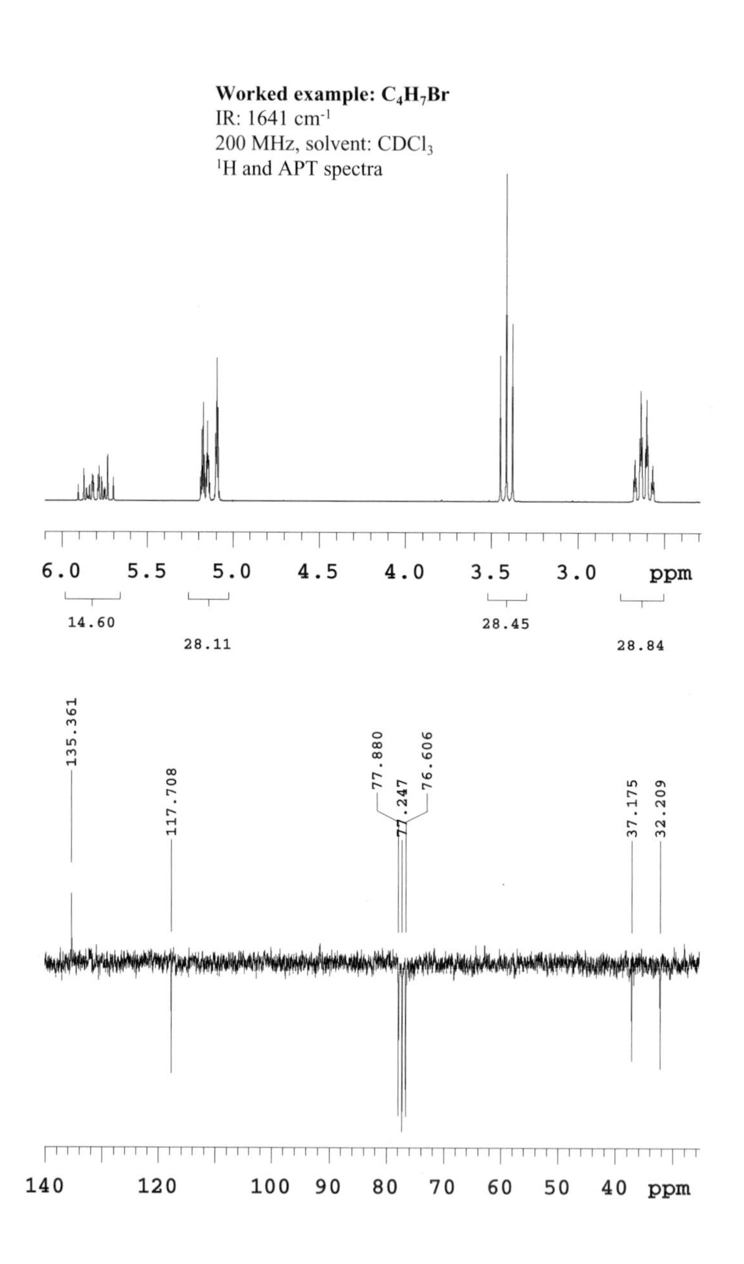

Worked example: C_4H_7Br
200 MHz, solvent: $CDCl_3$
H,H and C,H correlation

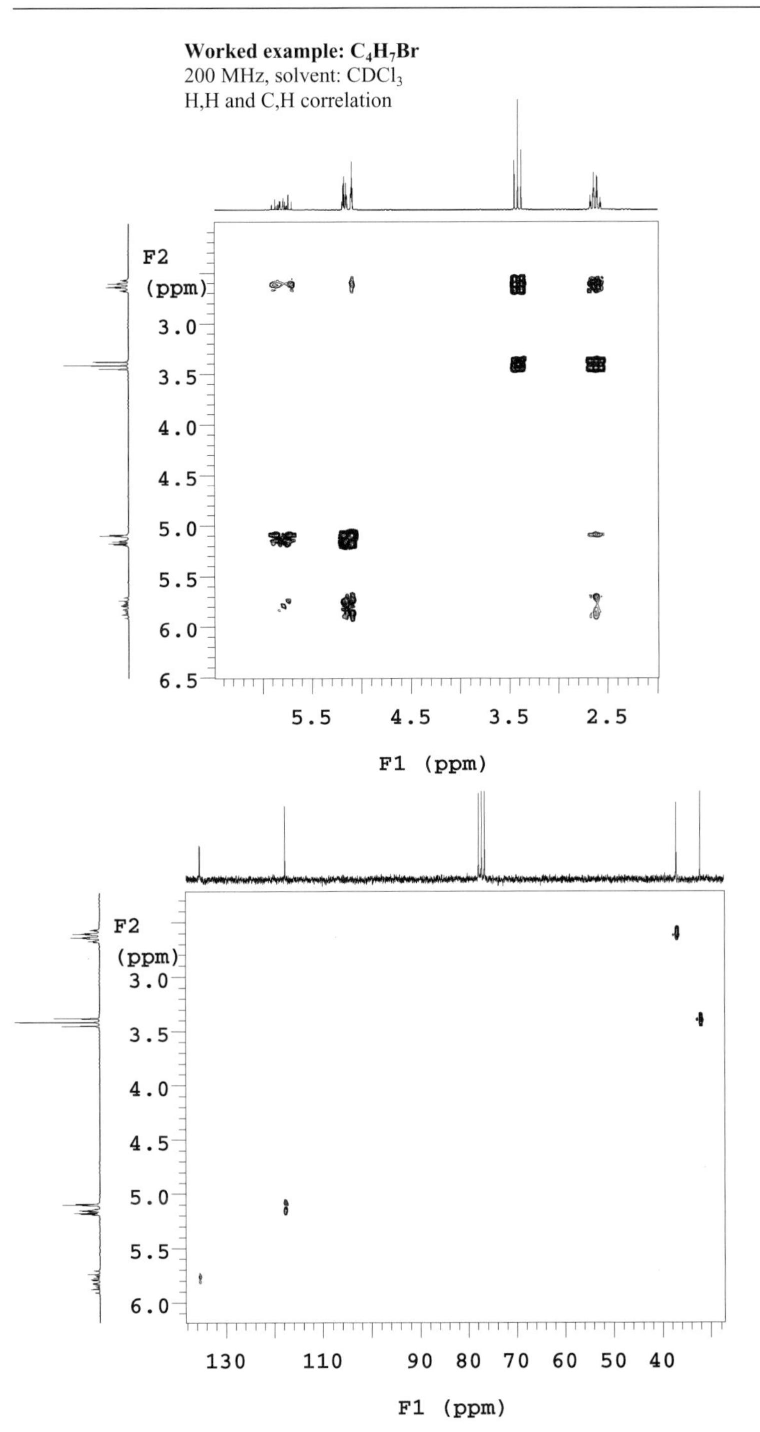

3
Problems

For solving the problems, please refer to the worked example (see p. 64) and the detailed data given in the headers of each problem.

Problem 1: C_4H_6NCl
IR: 2249 cm^{-1}
200 MHz, solvent: $CDCl_3$
^{1}H and ^{13}C spectra

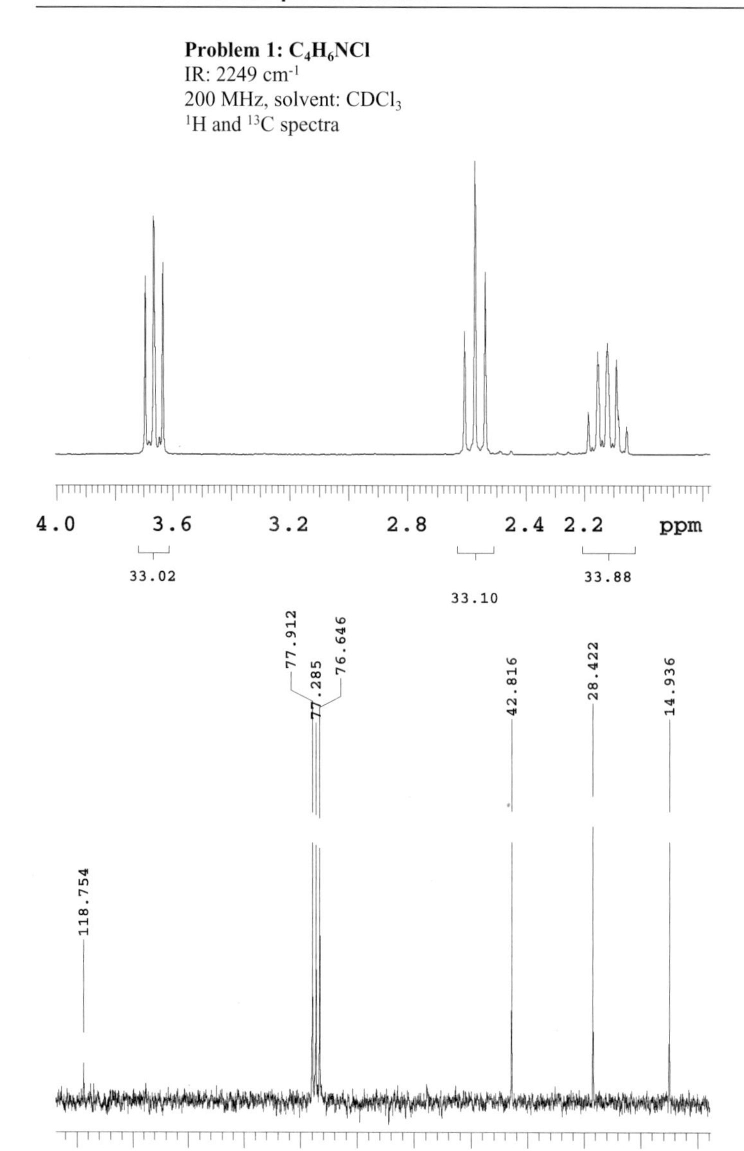

Problem 1: C_4H_6NCl
200 MHz, solvent: $CDCl_3$
H,H and C,H correlation

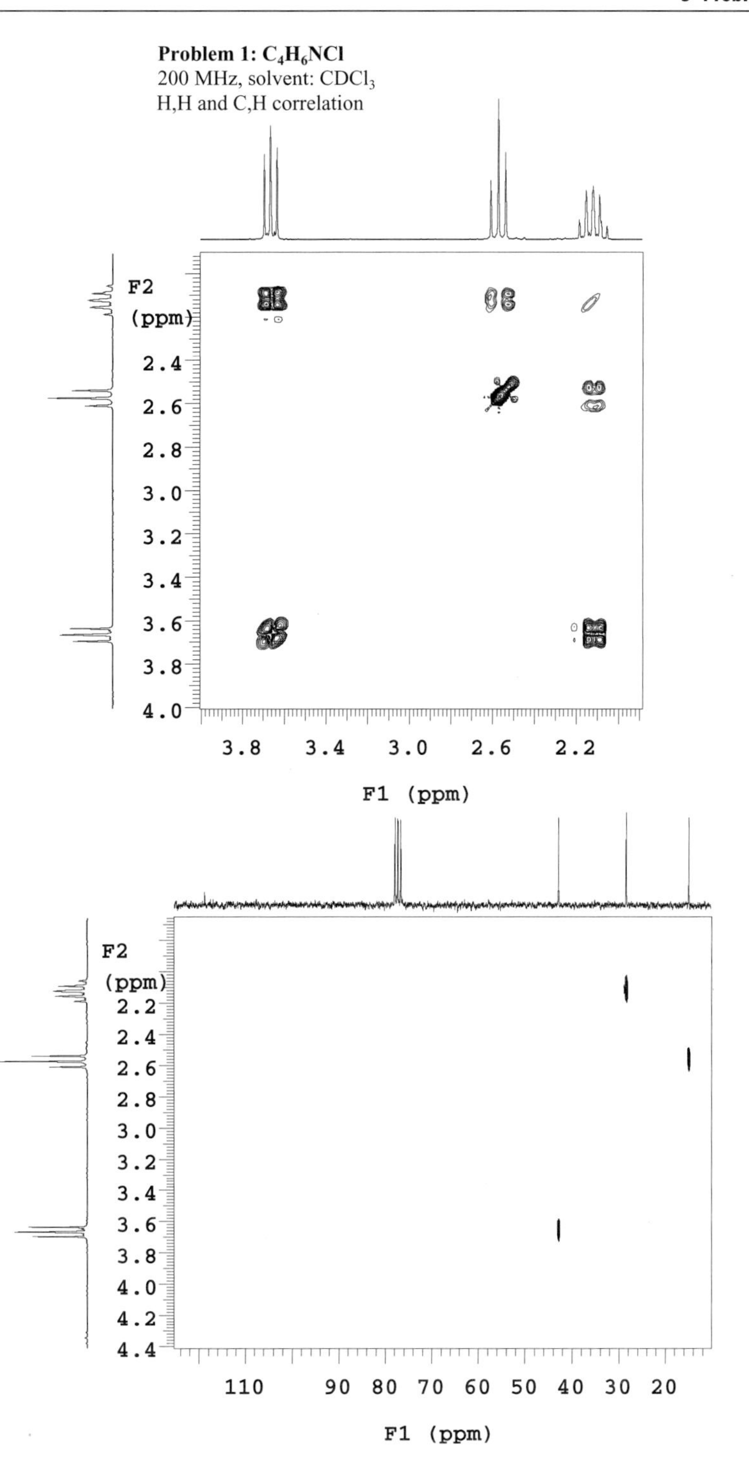

Problem 2: C_4H_6O
IR: 1654, 1691 cm^{-1}
200 MHz, solvent: $CDCl_3$
^1H and H, H correlation spectra

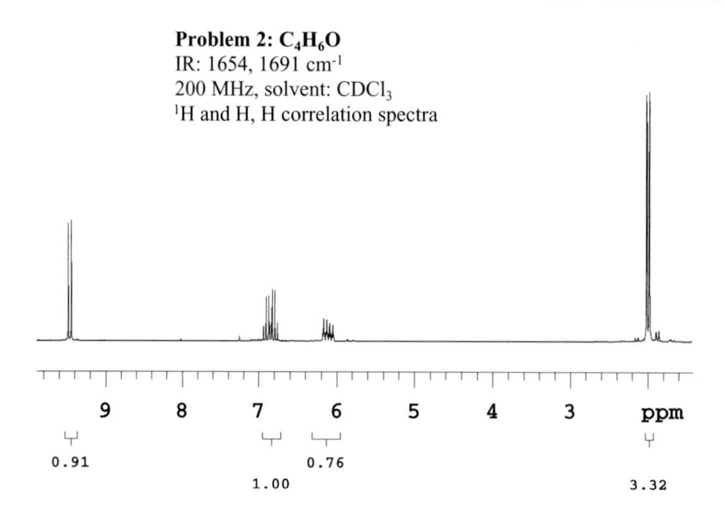

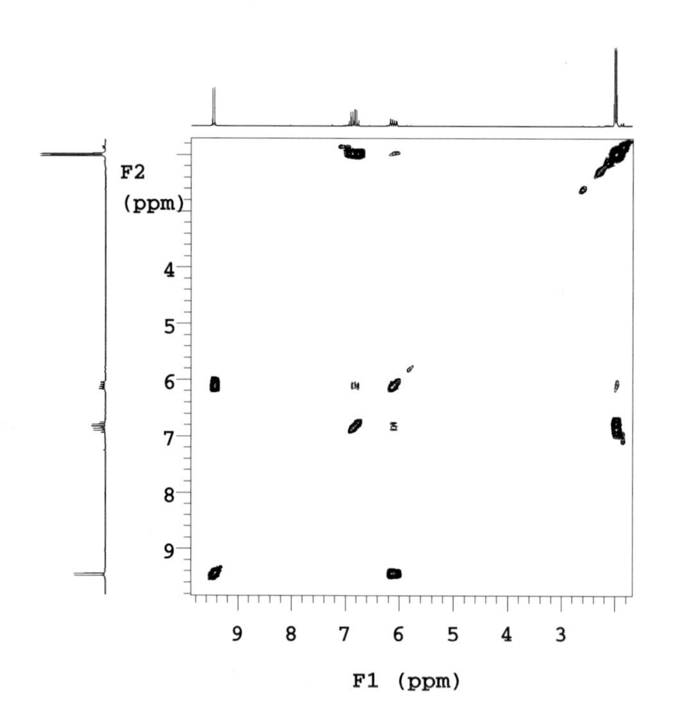

Problem 2: C_4H_6O
200 MHz, solvent: $CDCl_3$
^{13}C and DEPT spectra

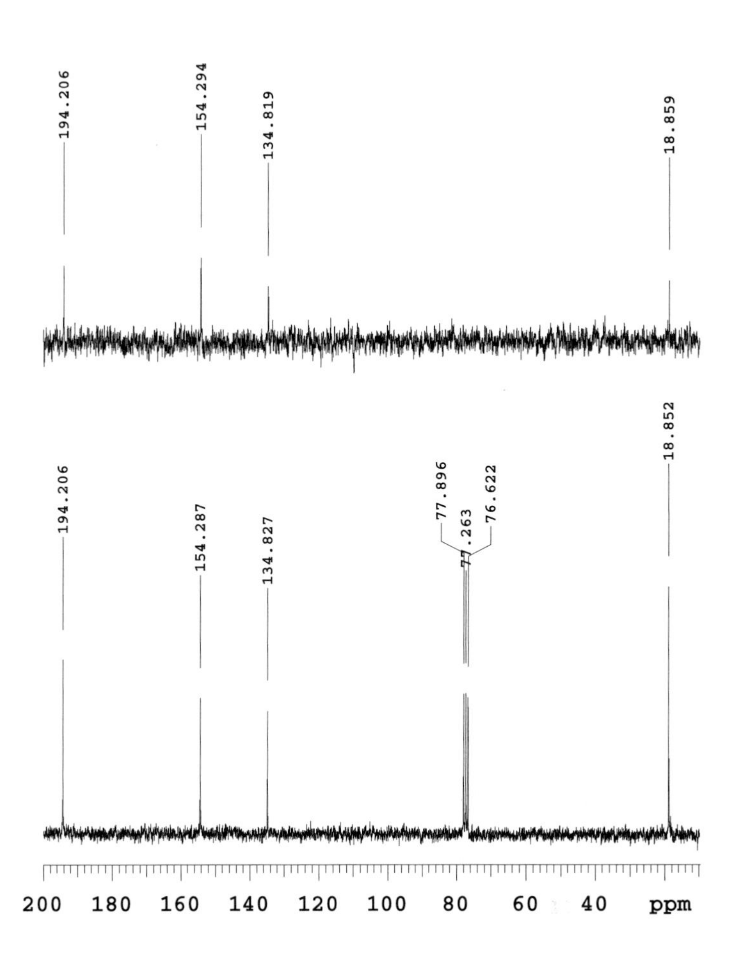

Problem 3: $C_4H_6O_3$
IR: 1730 cm^{-1} (broad)
200 MHz, solvent: $CDCl_3$
^1H and APT spectra

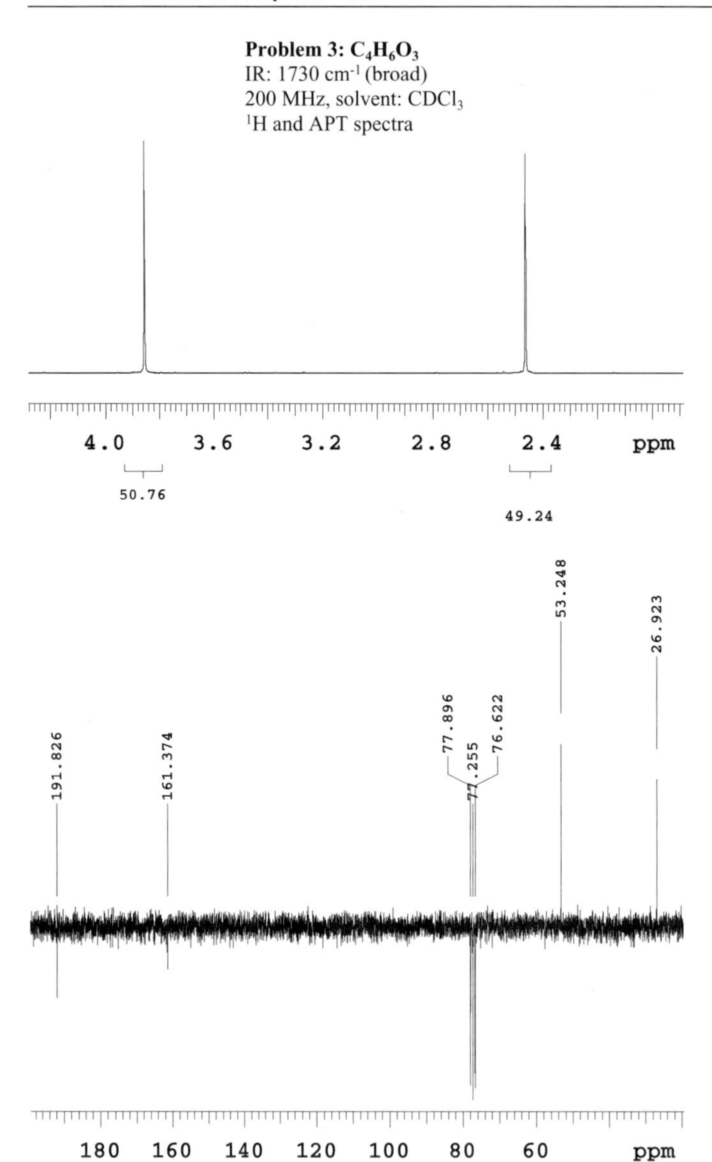

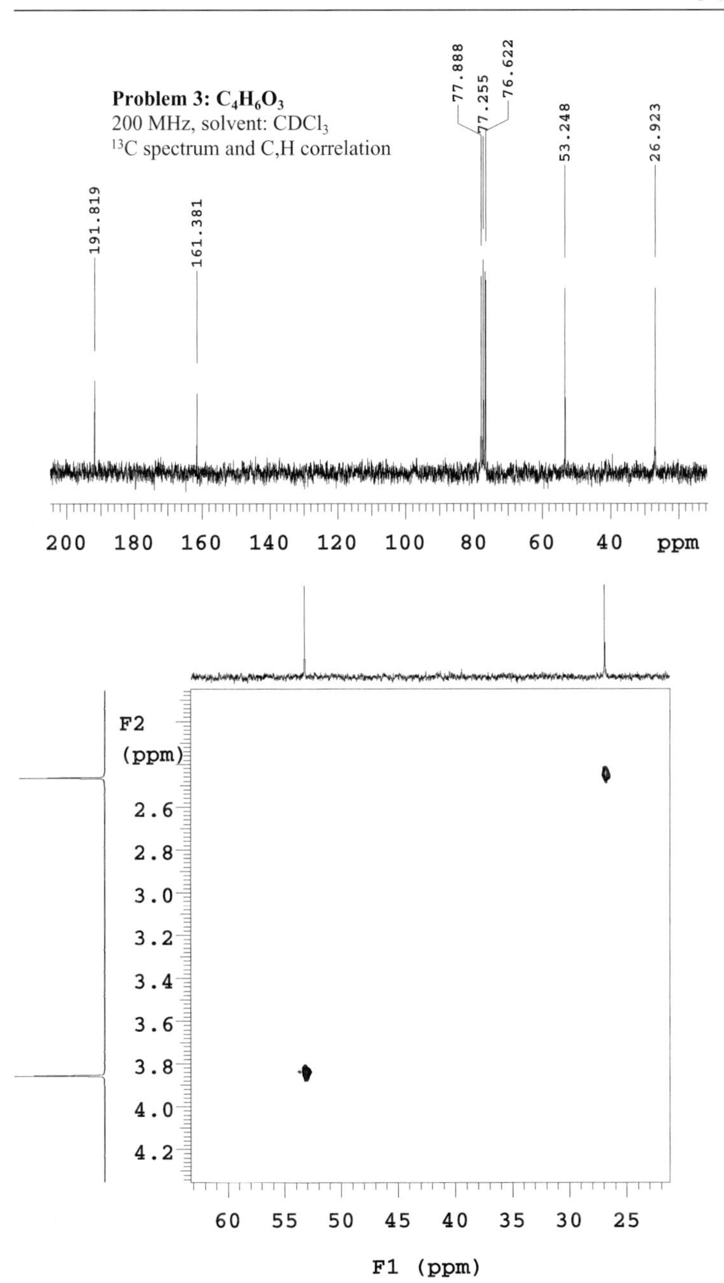
Problem 3: C4H6O3
200 MHz, solvent: CDCl3
13C spectrum and C,H correlation
191.819
161.381
77.888
77.255
76.622
53.248
26.923
200 180 160 140 120 100 80 60 40 ppm
F2
(ppm)
2.6
2.8
3.0
3.2
3.4
3.6
3.8
4.0
4.2
60 55 50 45 40 35 30 25
F1 (ppm)

Problem 4: $C_4H_{11}NO$
IR: 3350 cm^{-1} (broad)
200 MHz, solvent: $CDCl_3$
^1H and ^{13}C spectra

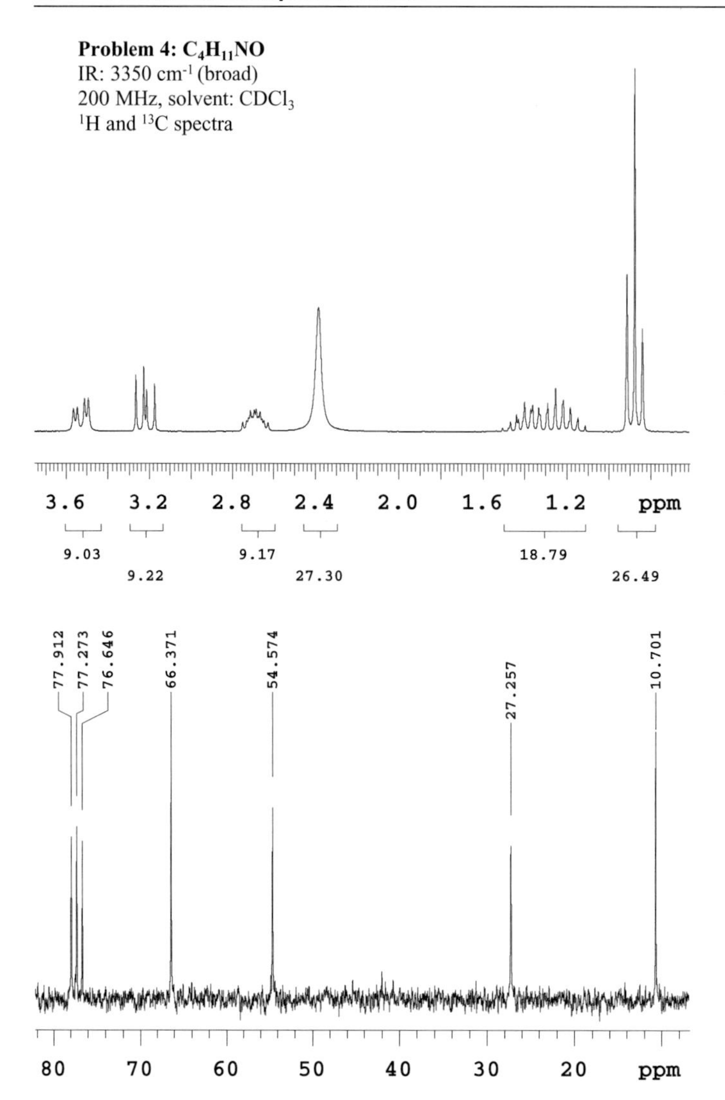

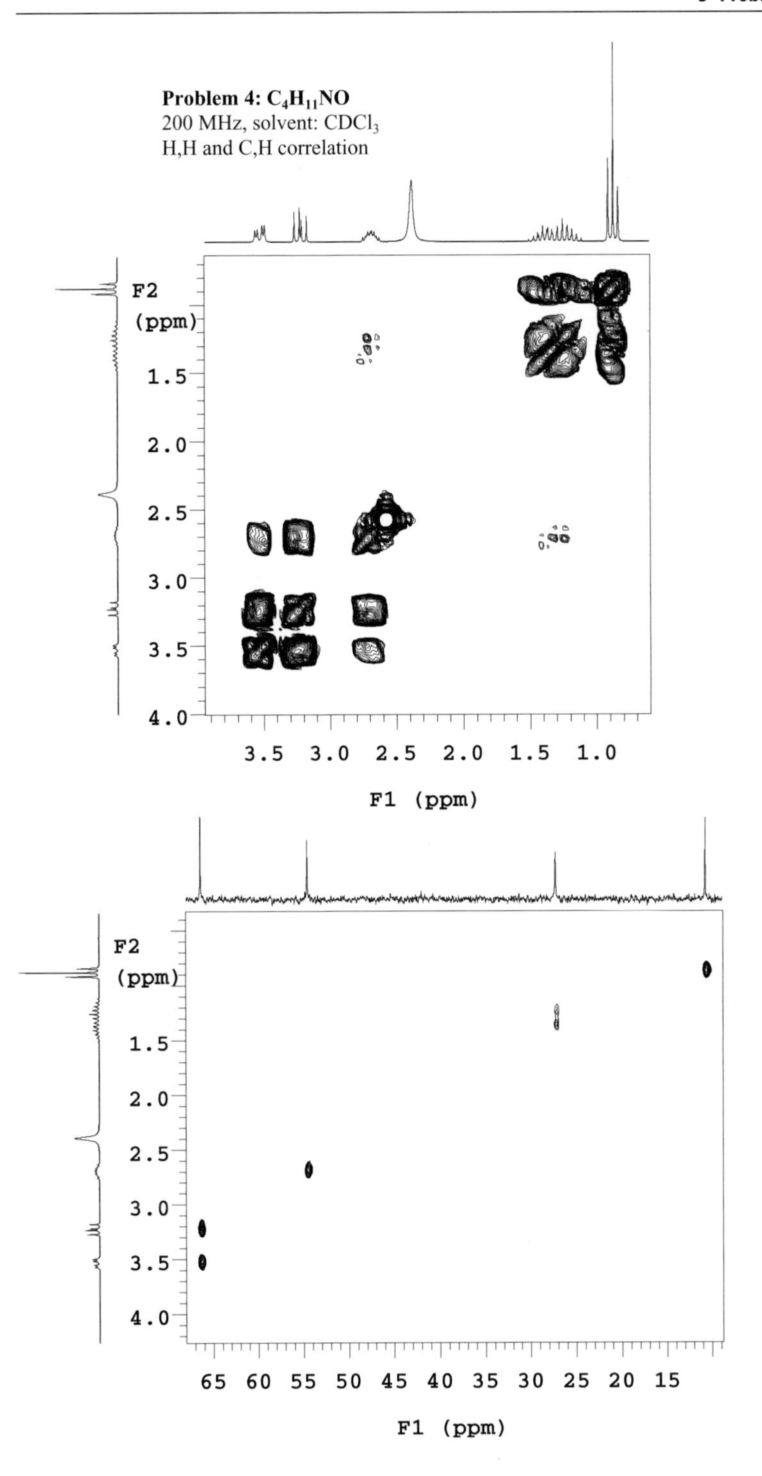

Problem 4: $C_4H_{11}NO$
200 MHz, solvent: $CDCl_3$
H,H and C,H correlation
F2
(ppm)
1.5
2.0
2.5
3.0
3.5
4.0
3.5 3.0 2.5 2.0 1.5 1.0
F1 (ppm)
F2
(ppm)
1.5
2.0
2.5
3.0
3.5
4.0
65 60 55 50 45 40 35 30 25 20 15
F1 (ppm)

Problem 5: $C_5H_3N_2O_2Cl$
IR: 1355, 1562 cm^{-1}
200 MHz, solvent: $CDCl_3$
^1H and APT spectra

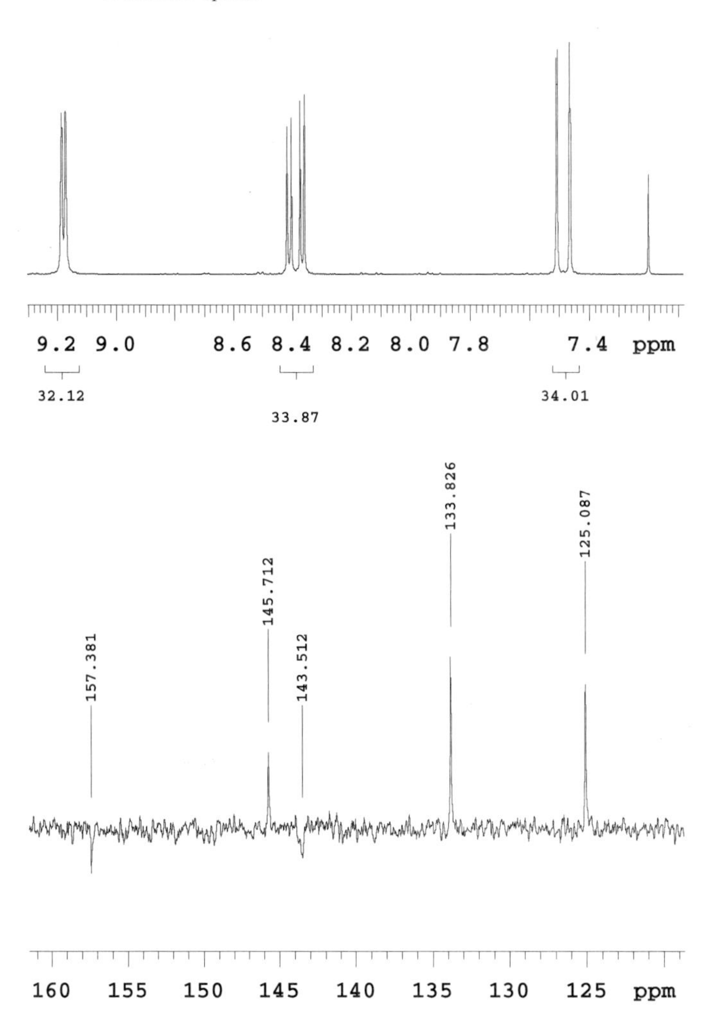

Problem 5: $C_5H^3N_2O_2Cl$
200 MHz, solvent: $CDCl_3$
H,H and C,H correlation

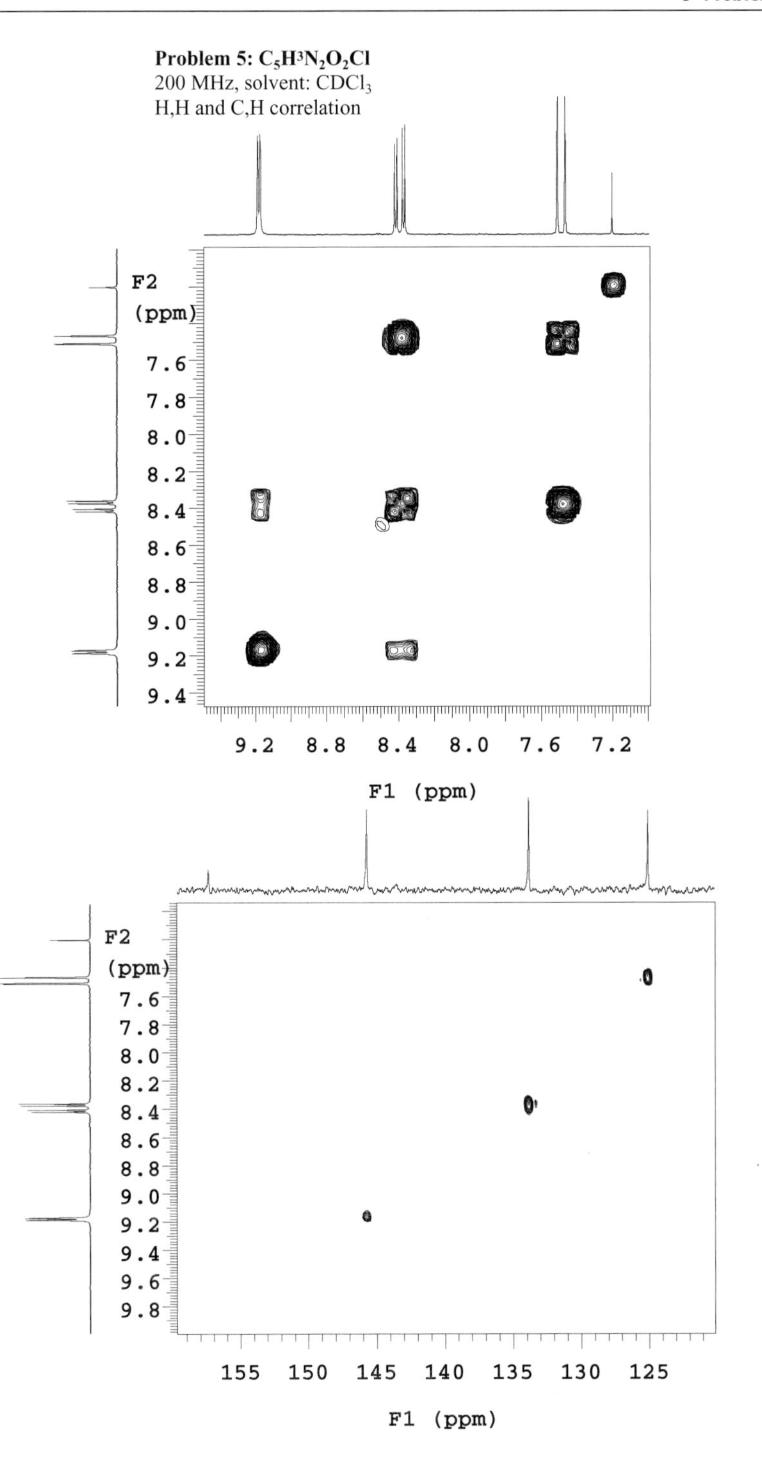

Problem 6: $C_5H_6N_2$

IR: 3296, 3360 cm^{-1}

500 MHz, solvent: $CDCl_3$

^1H and APT spectra

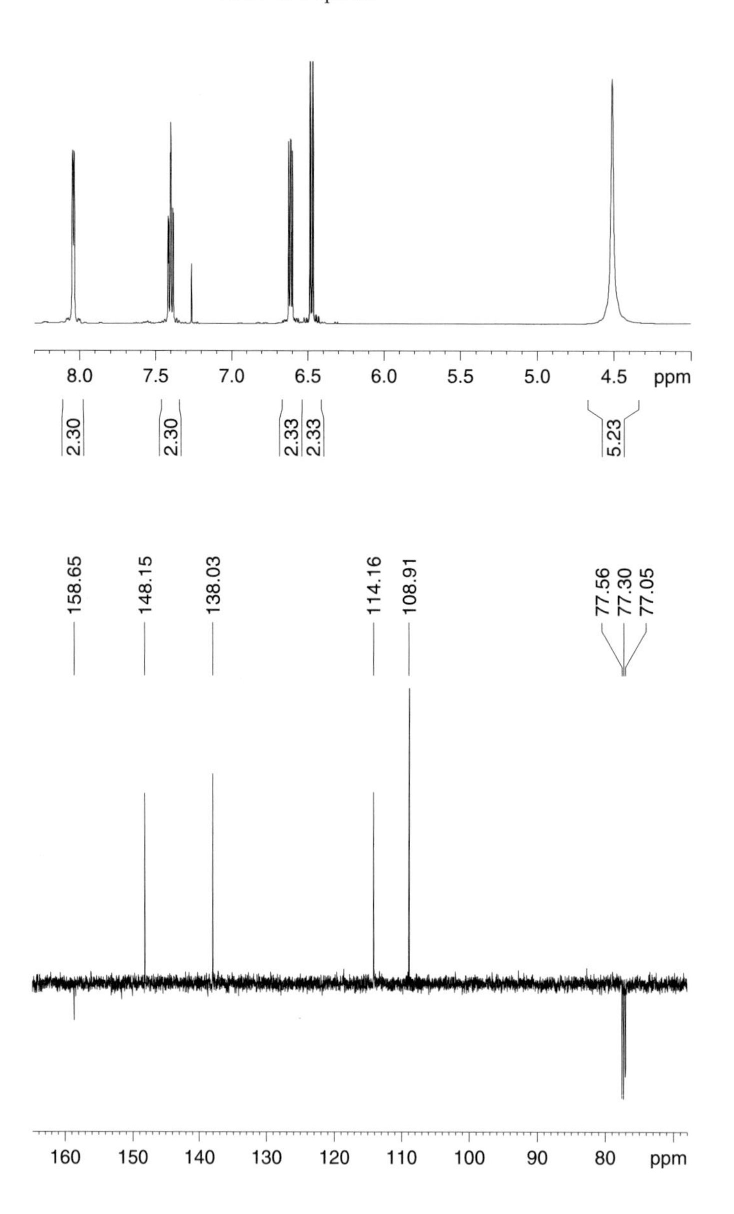

Problem 6: $C_5H_6N_2$
500 MHz, solvent: $CDCl_3$
H,H and C,H correlation

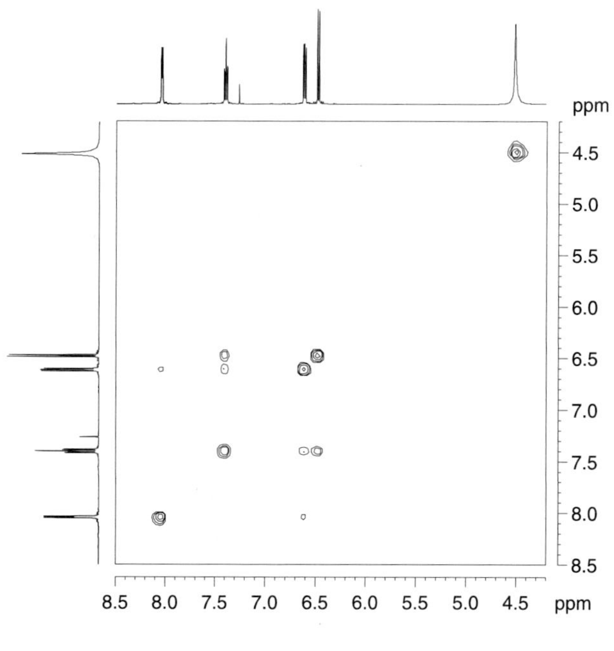

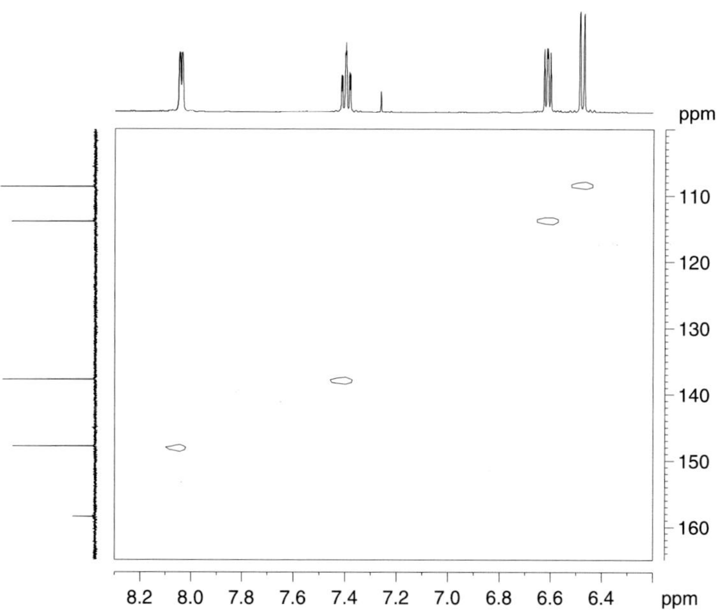

Problem 7: C_6H_5OF

^{19}F NMR: δ =-125 ppm

IR: 3215 cm^{-1}

600 MHz, solvent: $CDCl_3$

^1H, ^{13}C and DEPT spectra

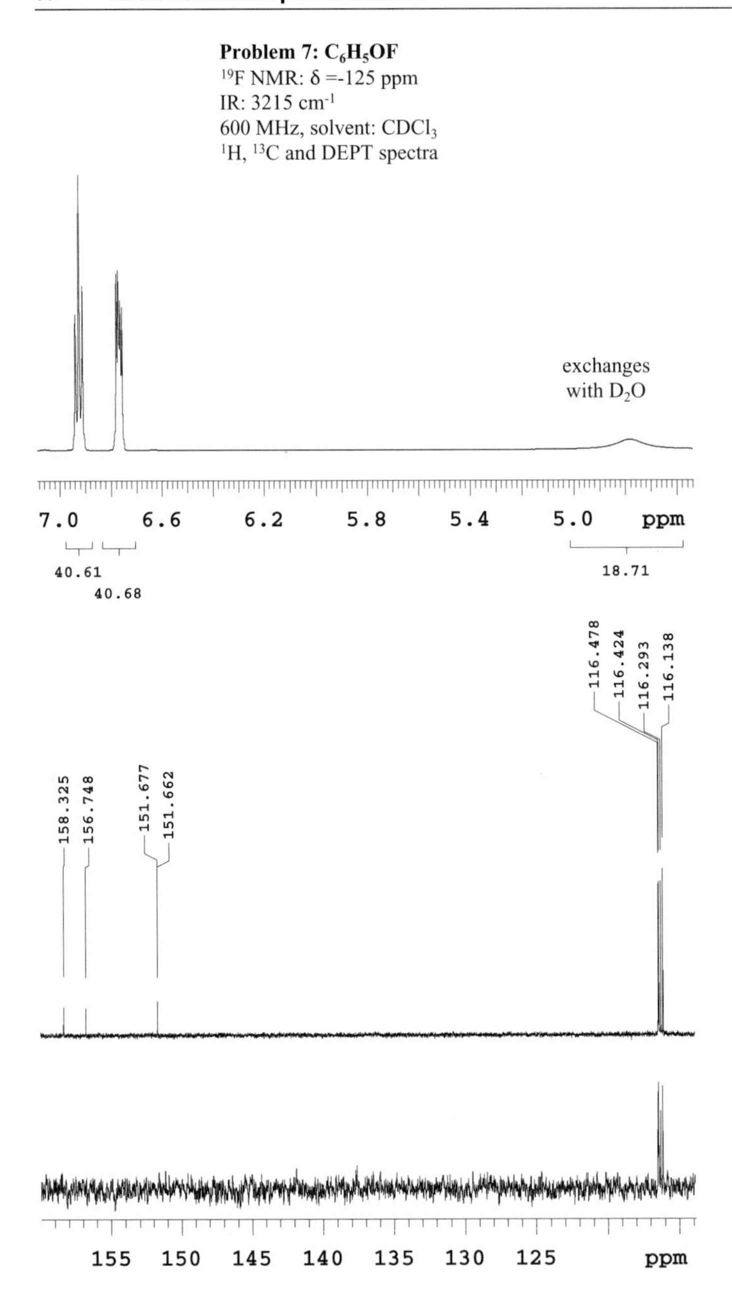

Problem 7: C_6H_5OF
600 MHz, solvent: $CDCl_3$
^{13}C spectra (expansion)
scale in ppm, peak frequency in Hz

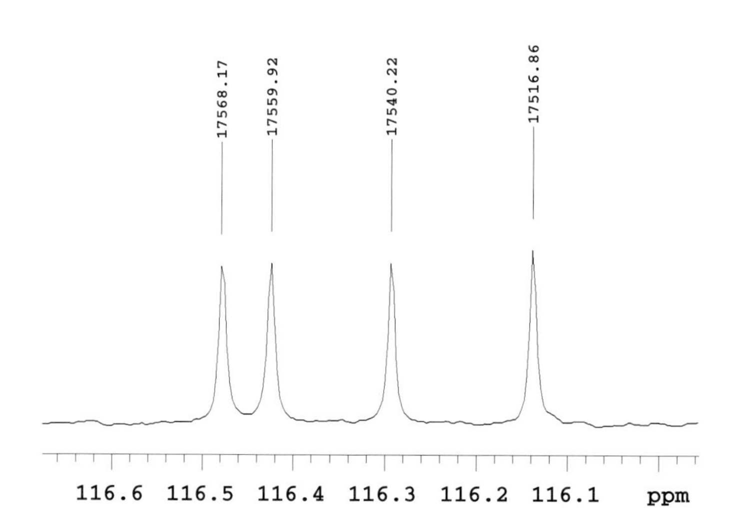

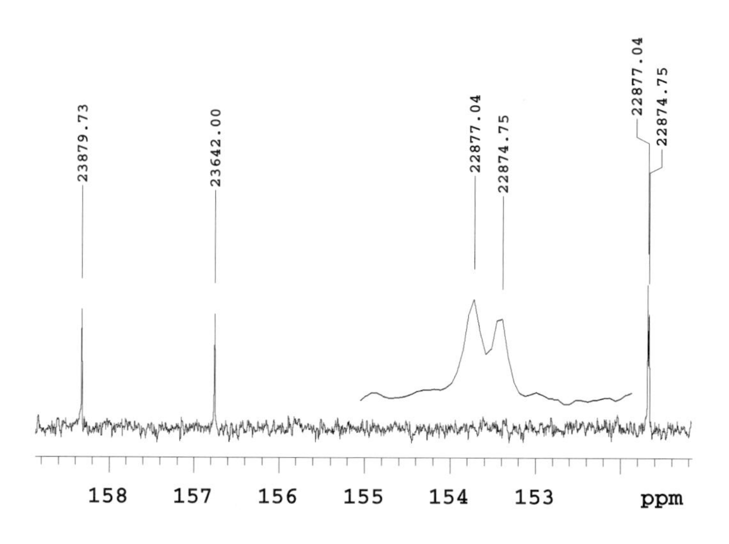

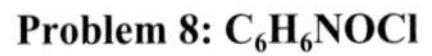

Problem 8: C_6H_6NOCl

IR: no bands characteristic of functional groups
200 MHz, solvent: $CDCl_3$
^1H, ^{13}C and DEPT spectra

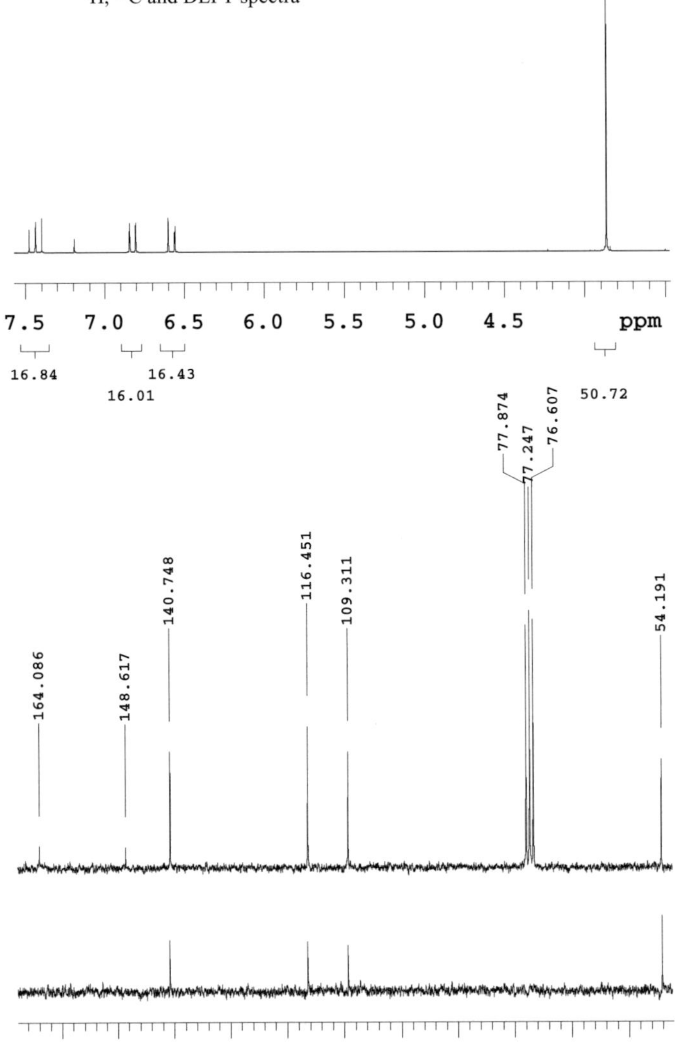

Problem 8: C_6H_6NOCl
200 MHz, solvent: $CDCl_3$
H,H and C,H correlation

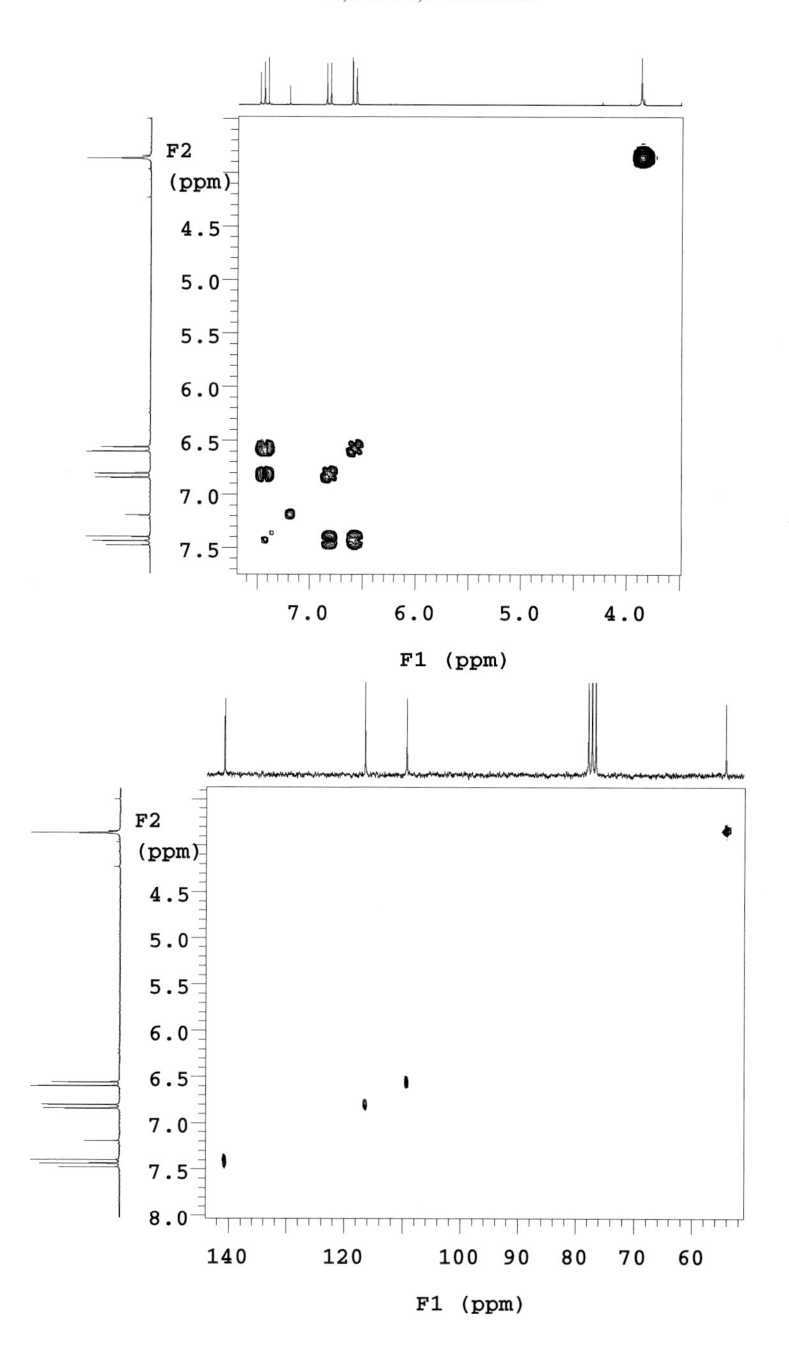

Problem 9: C_6H_7NO
IR: 3166, 3304, 3360 cm^{-1}
600 MHz, solvent: DMSO-d6
^1H and ^{13}C spectra

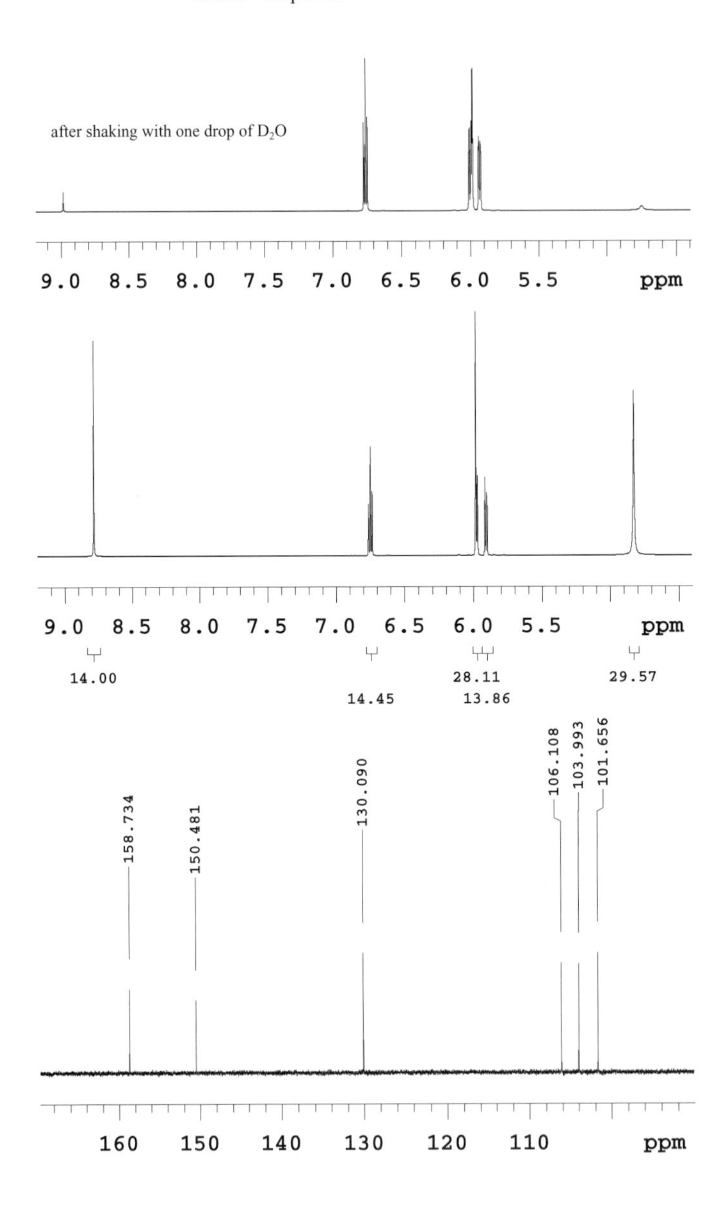

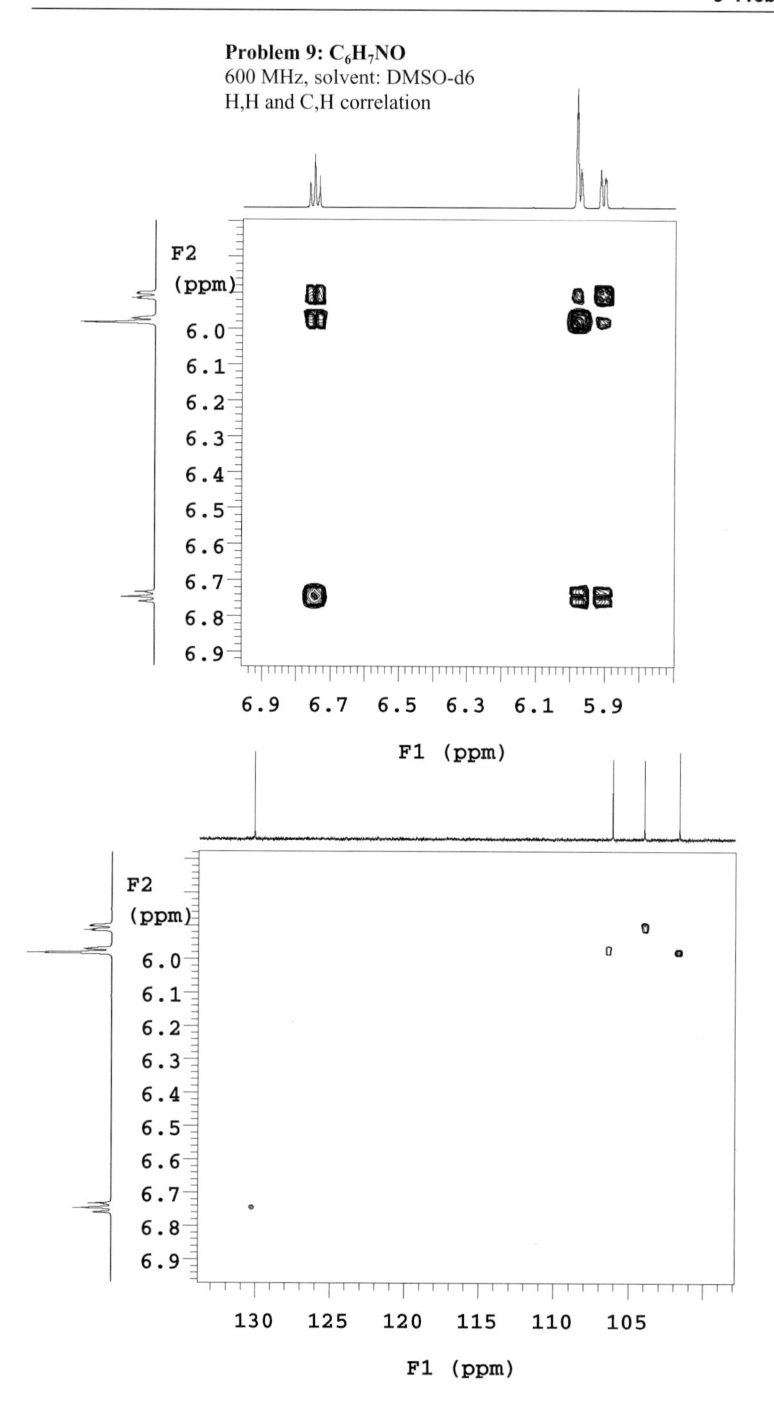

Problem 9: C6H7NO
600 MHz, solvent: DMSO-d6
H,H and C,H correlation
F2
(ppm)
6.0
6.1
6.2
6.3
6.4
6.5
6.6
6.7
6.8
6.9
6.9 6.7 6.5 6.3 6.1 5.9
F1 (ppm)
F2
(ppm)
6.0
6.1
6.2
6.3
6.4
6.5
6.6
6.7
6.8
6.9
130 125 120 115 110 105
F1 (ppm)

Problem 10: $C_7H_4N_2O_6$
IR: 1348, 1545, 1703, 3093 cm^{-1}
200 MHz, solvent: DMSO-d6
^1H and ^{13}C spectra
^1H expansion: peak frequency in Hz

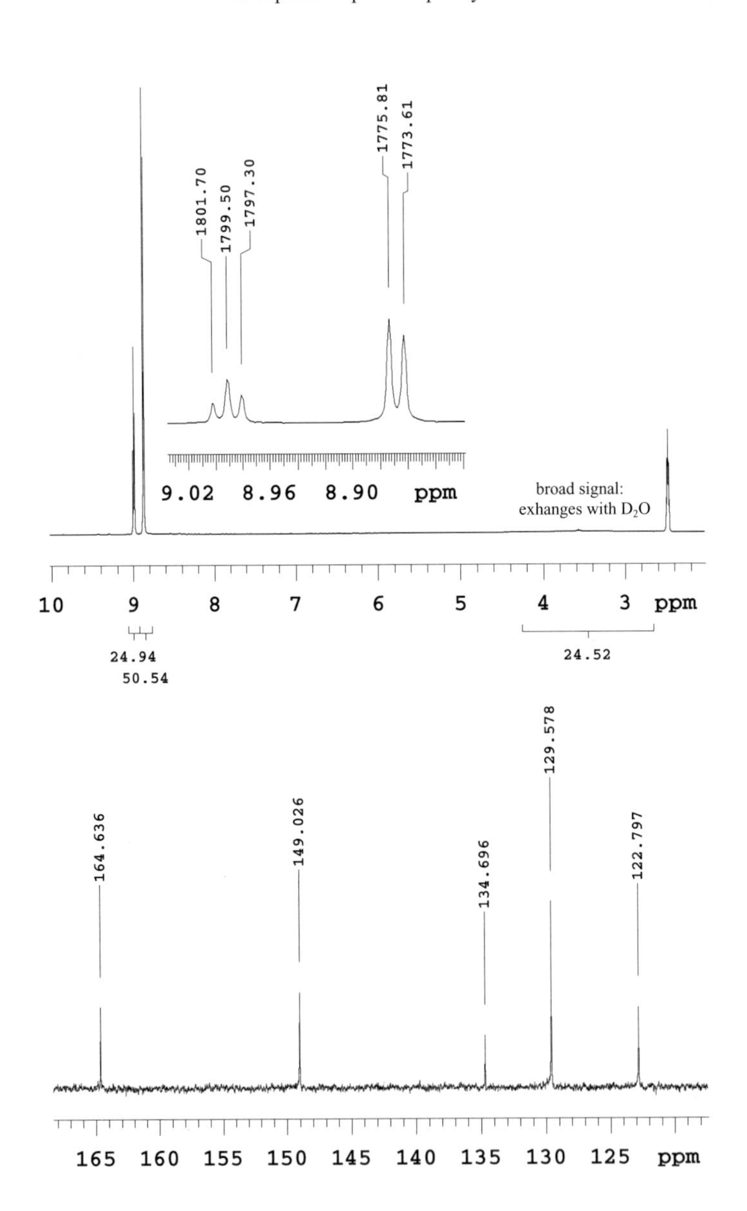

Problem 10: $C_7H_4N_2O_6$
200 MHz, solvent: DMSO-d6
APT and DEPT spectra

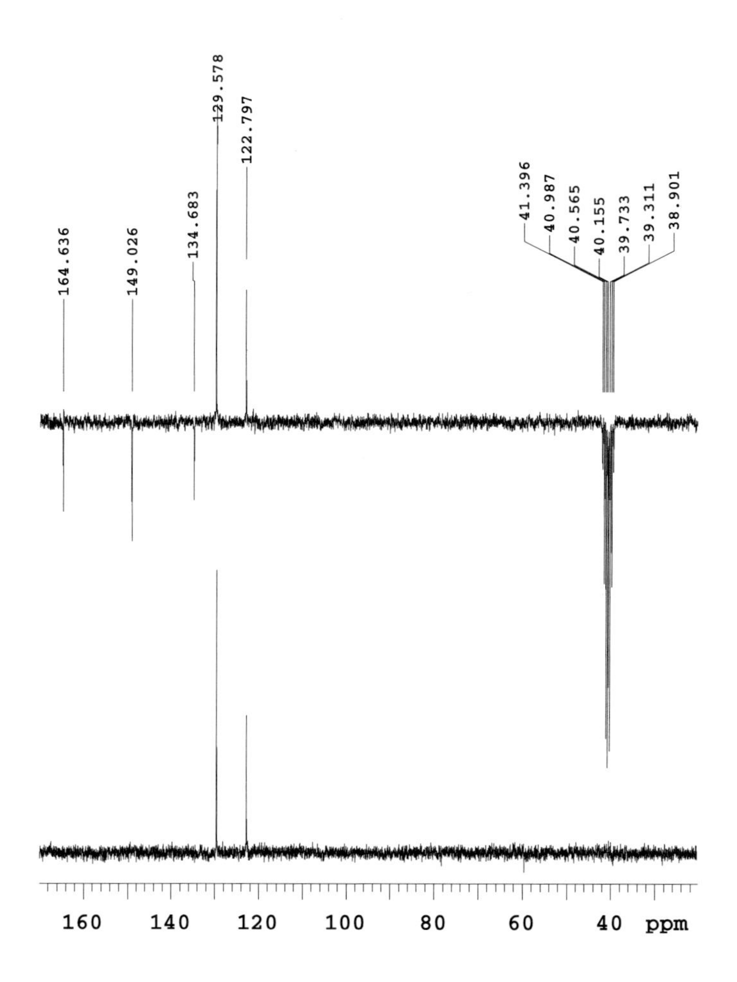

Problem 11: C_7H_8NCl
IR: 3355, 3430 cm^{-1}
500 MHz, solvent: $CDCl_3$
^1H, ^{13}C and DEPT spectra

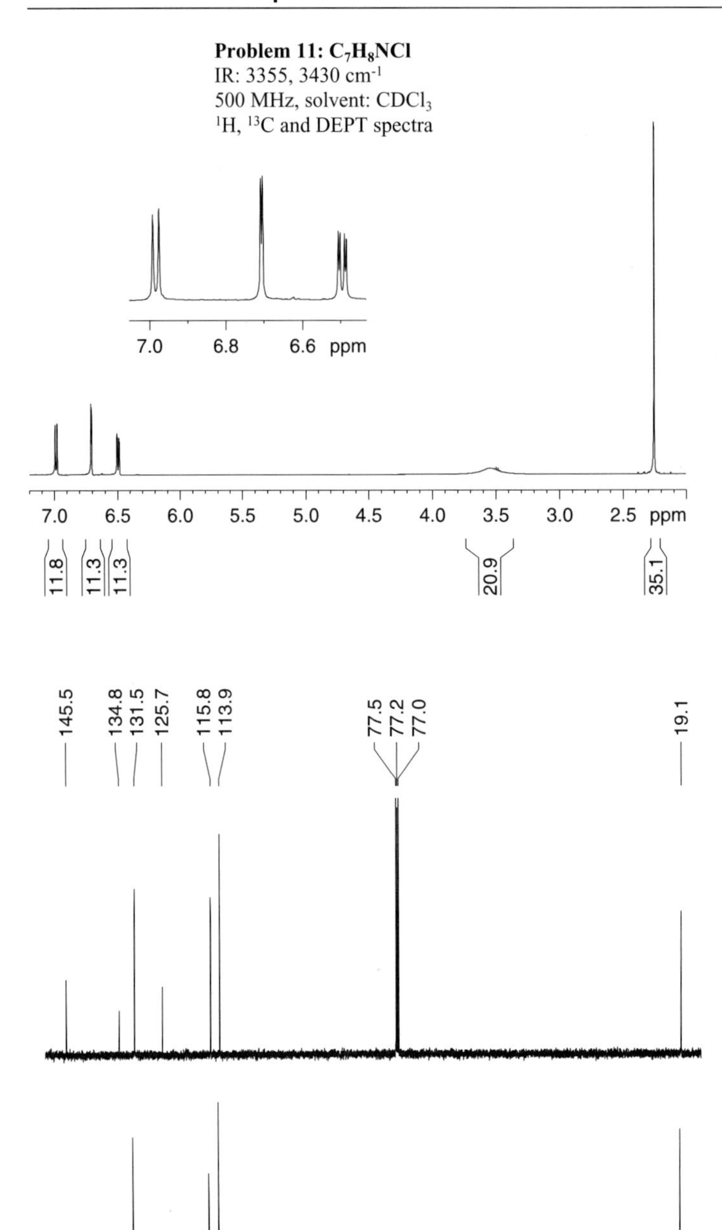

Problem 11: C_7H_8NCl
solvent: $CDCl_3$
H,H correlation, 200 MHz
C,H correlation, 500 MHz

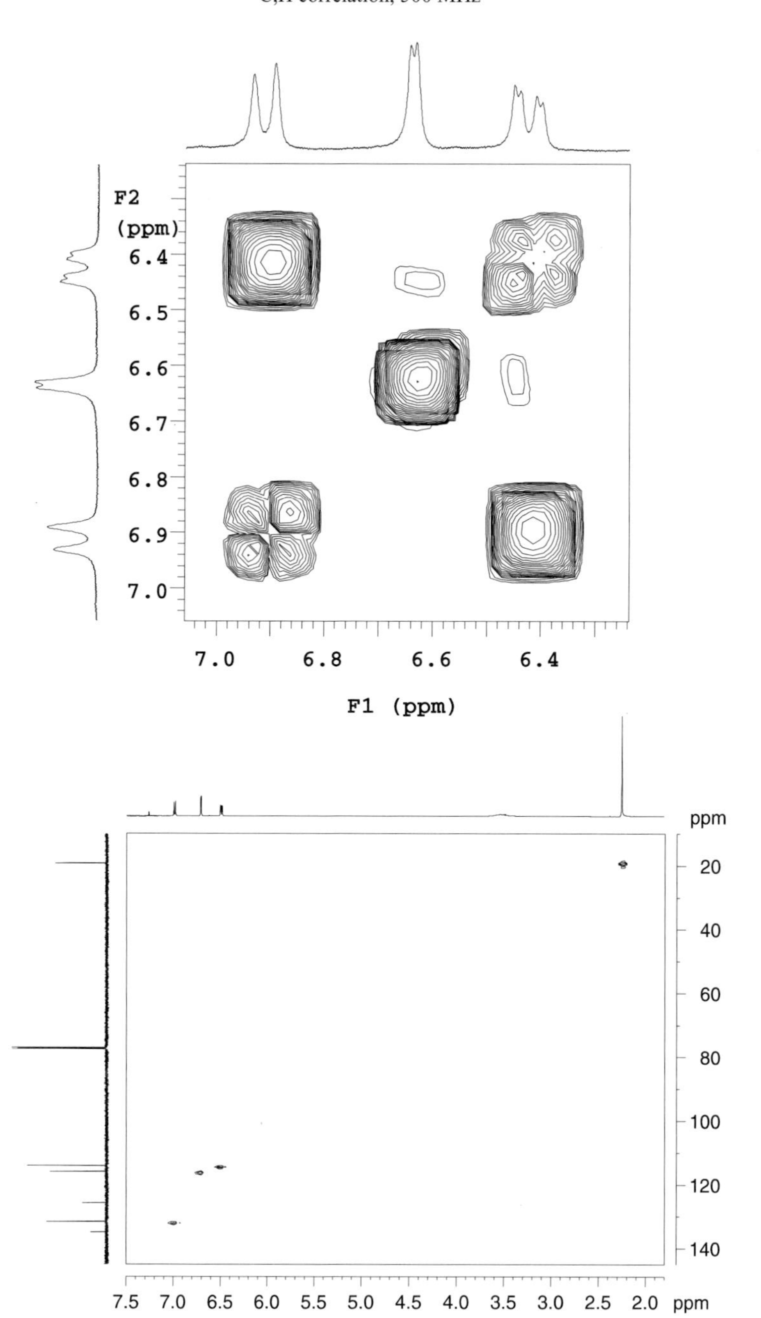

Problem 12: $C_7H_{10}O_5$

IR: 1720 (strong), 2930-2970 (broad) cm^{-1}

200 MHz, solvent: DMSO-d6 (with small amount of water: $\delta = 3.4$ ppm broad)

^1H spectra with expansion

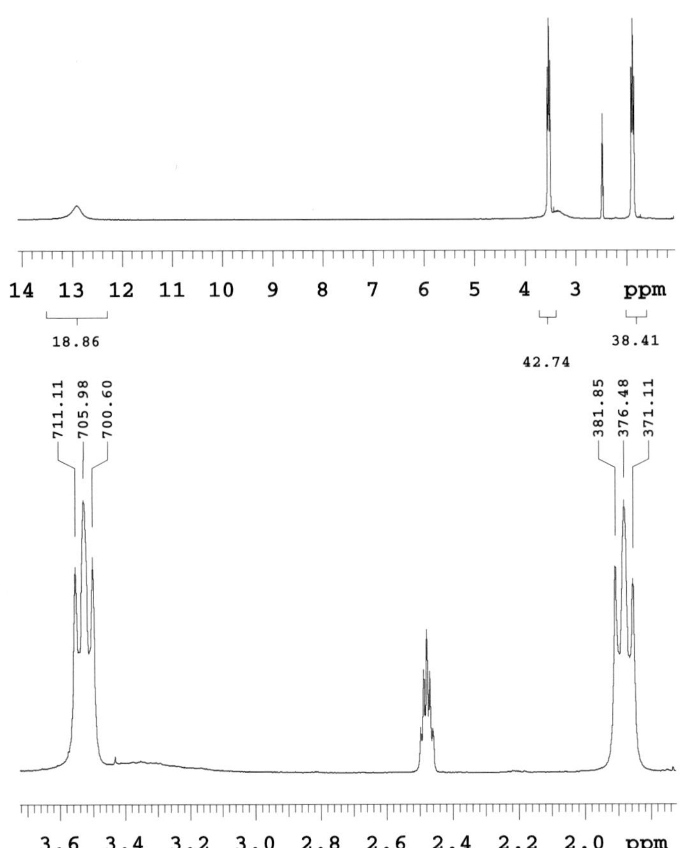

Problem 12: $C_7H_{10}O_5$
200 MHz, solvent: DMSO-d6
^{13}C and DEPT spectra

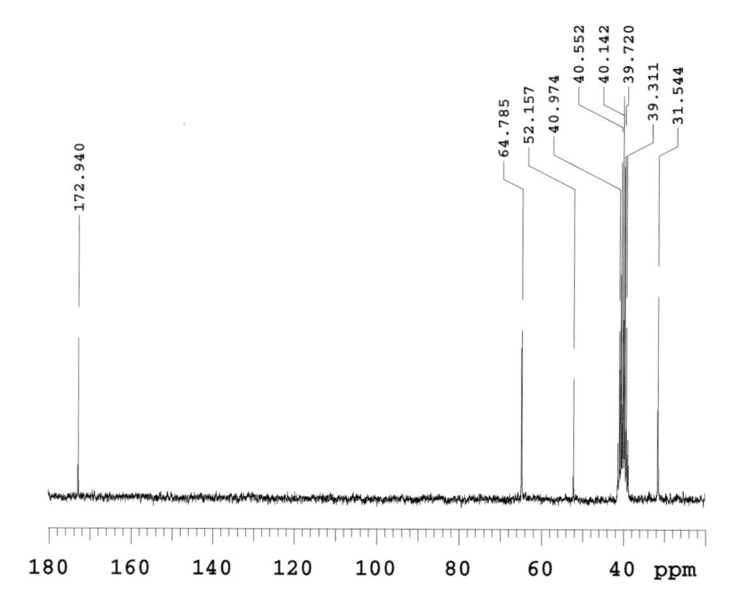

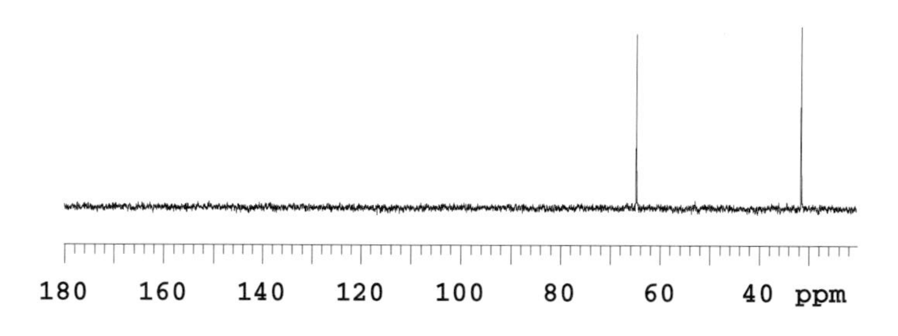

Problem 13: C_8H_7OBr

IR: 1641 cm^{-1}

200 and 600 MHz, solvent: $CDCl_3$

^{1}H spectra and H,H correlation

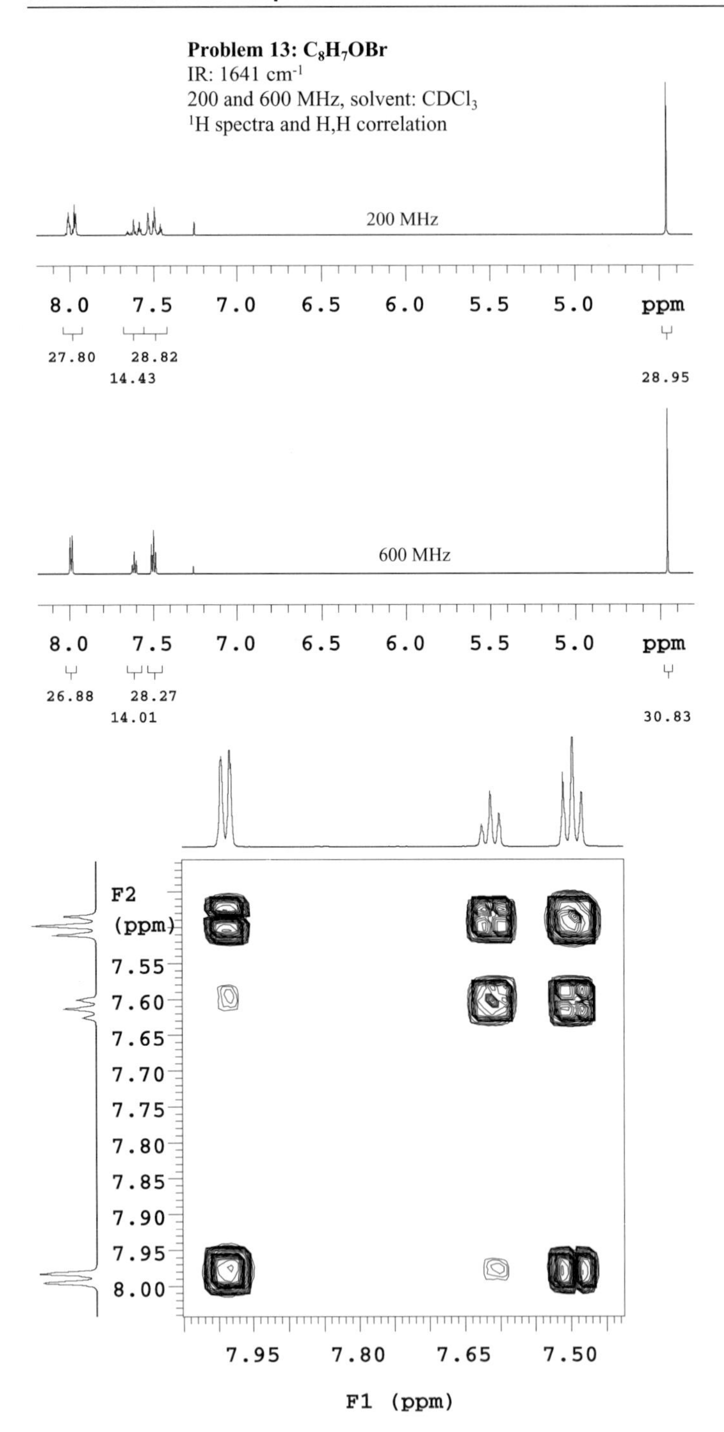

Problem 13: C_8H_7OBr
200 MHz, solvent: $CDCl_3$
^{13}C spectra and C,H correlation

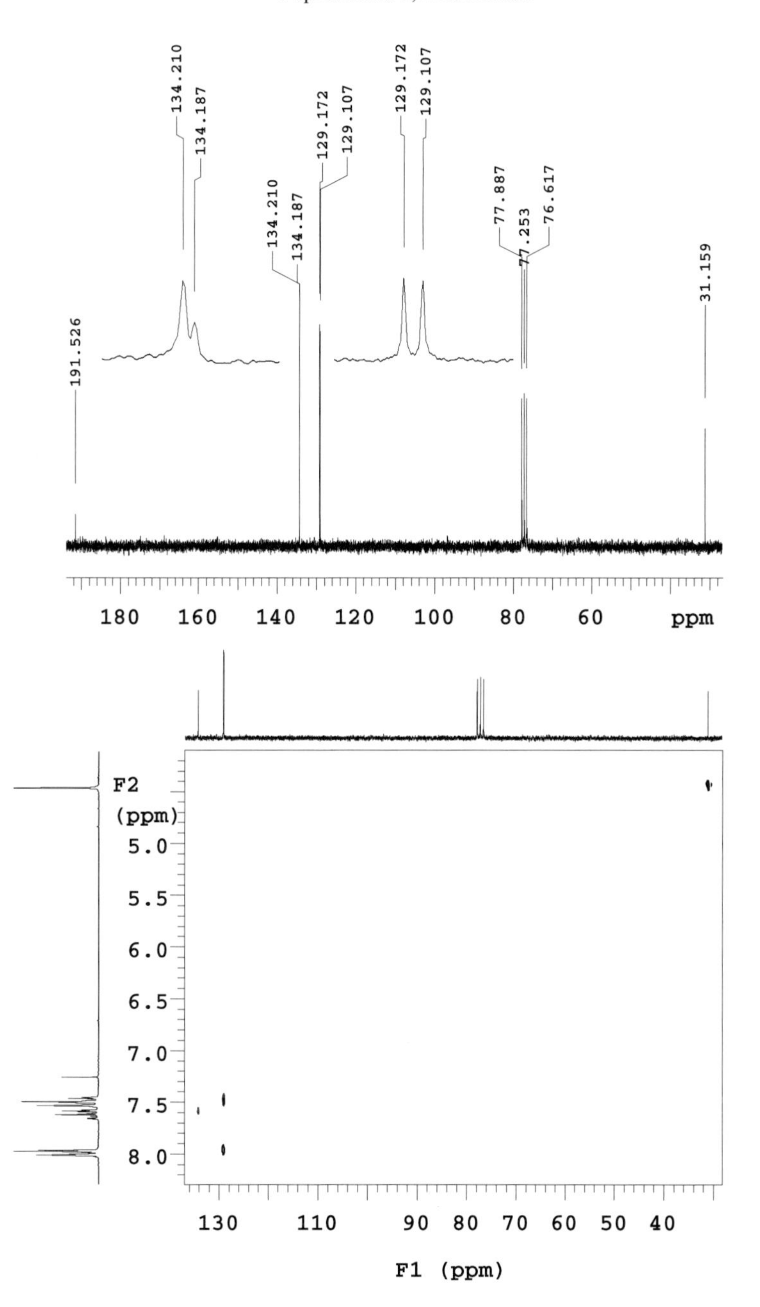

Problem 14: $C_8H_{10}N_2O$
IR: 1681, 3093, 3159, 3279 cm^{-1}
200 MHz, solvent: $CDCl_3$ (top) and DMSO-d6 (below)
^1H spectra

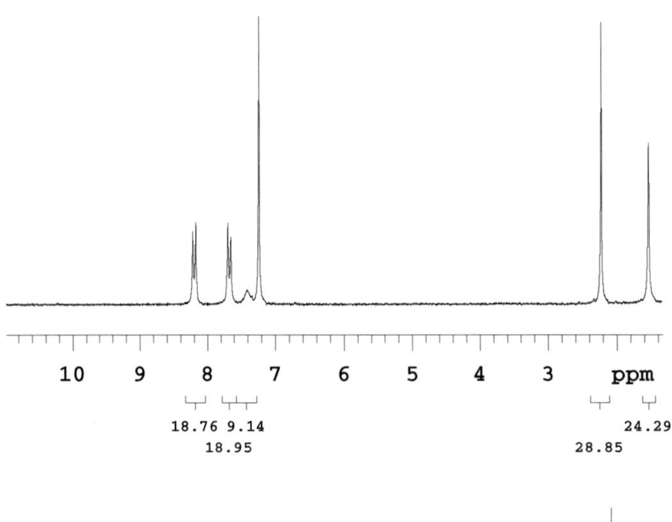

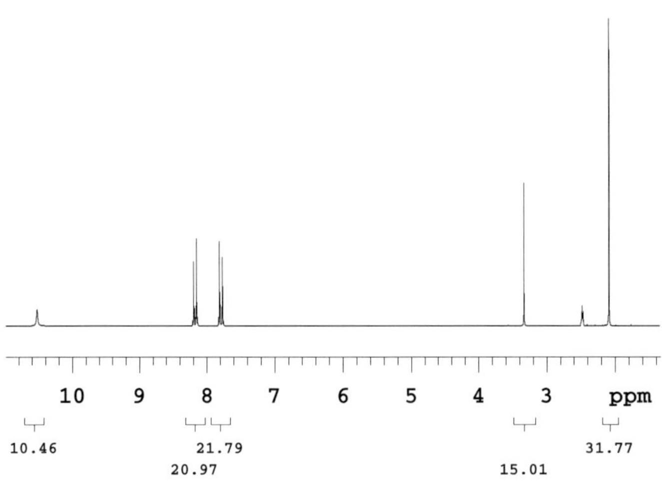

Problem 14: $C_8H_{10}N_2O$
200 MHz, solvent: DMSO-d6
^{13}C and DEPT spectra

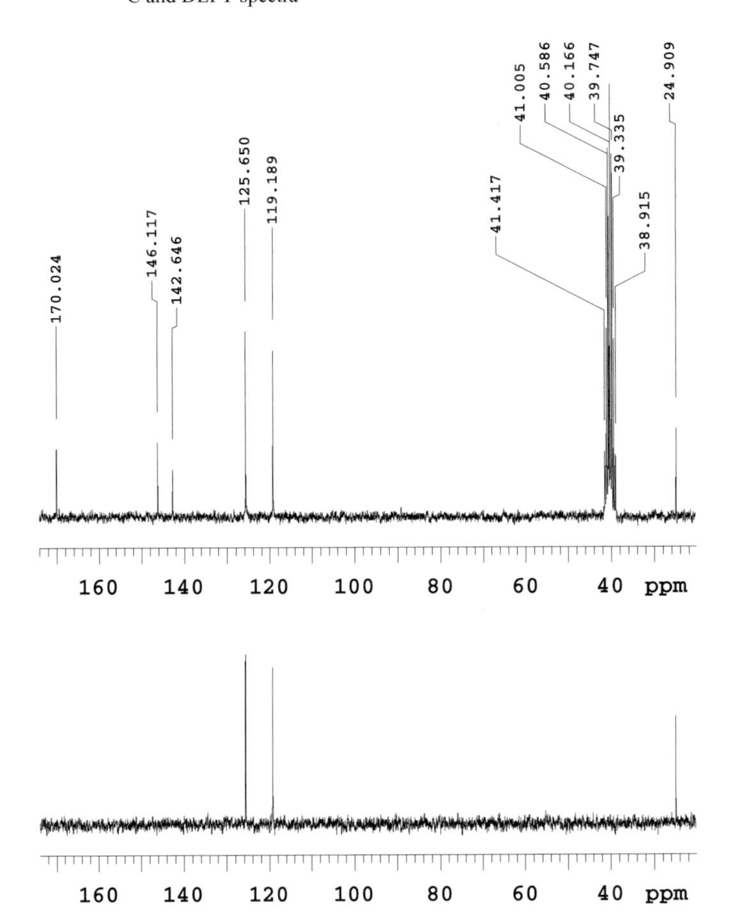

Problem 15: $C_8H_{10}O_2$
IR: 3365 (very broad) cm^{-1}
500 MHz, solvent: $CDCl_3$
^1H, ^{13}C and DEPT spectra

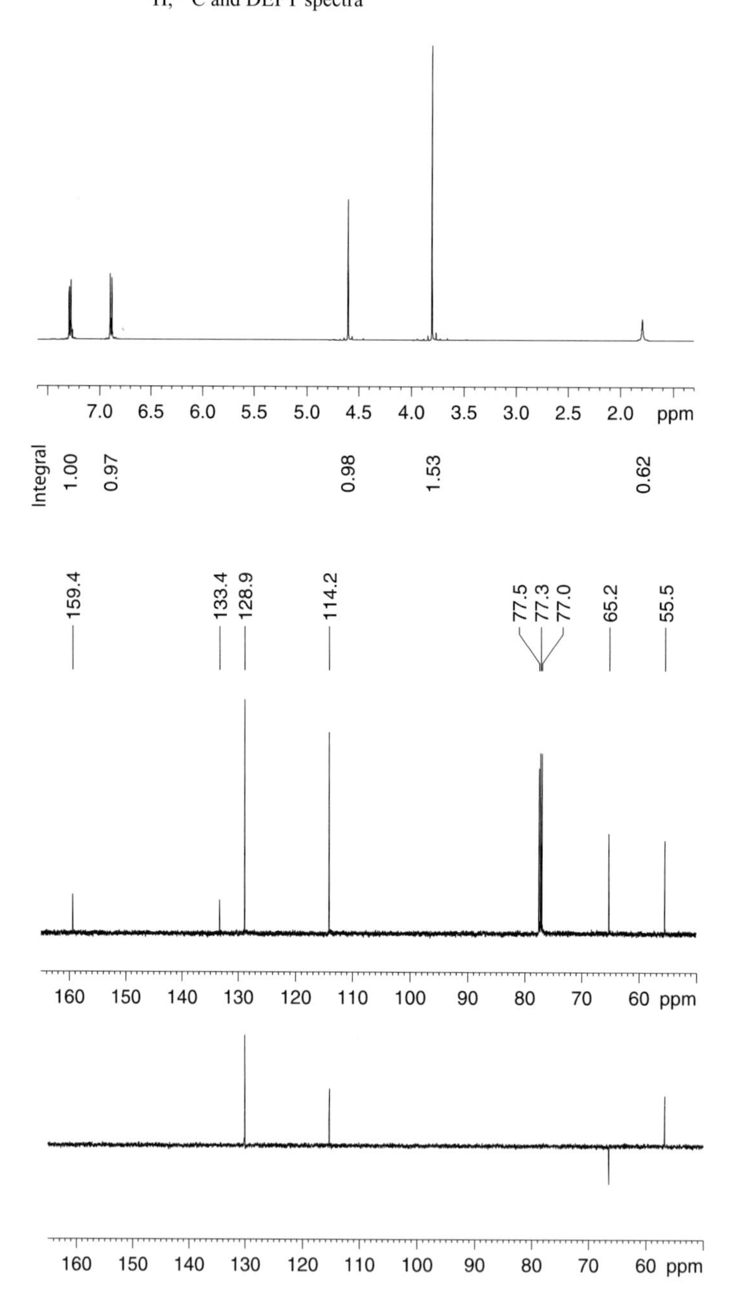

Problem 15: $C_8H_{10}O_2$
500 MHz, solvent: $CDCl_3$
H,H and C,H correlation

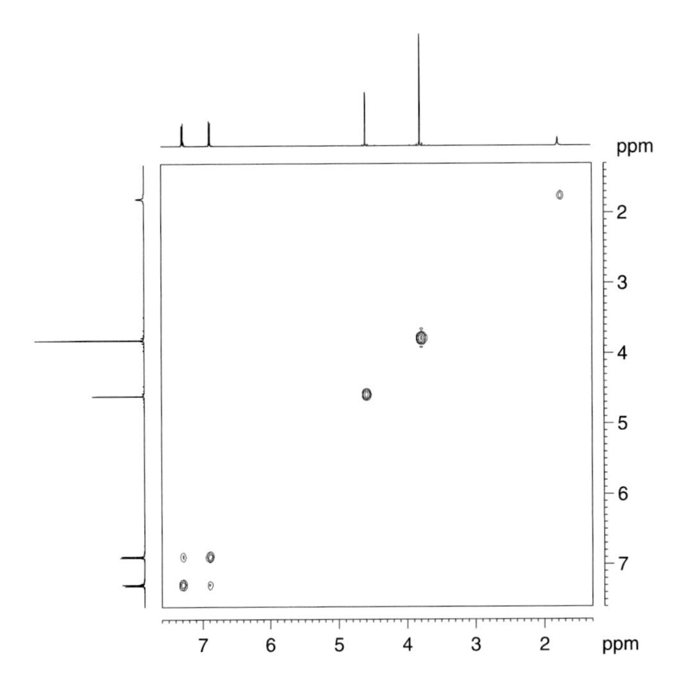

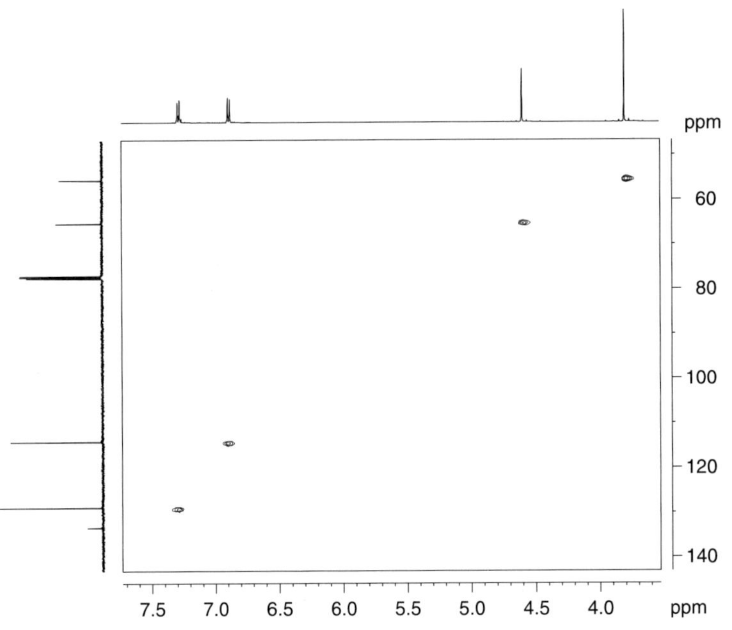

Problem 16: $C_8H_{11}N$
IR: 1637, 2214 cm^{-1}
200 MHz, solvent: $CDCl_3$
^1H and APT spectra

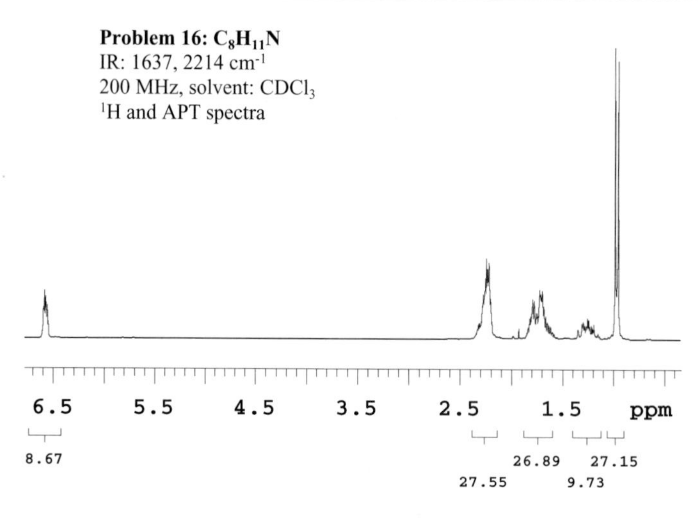

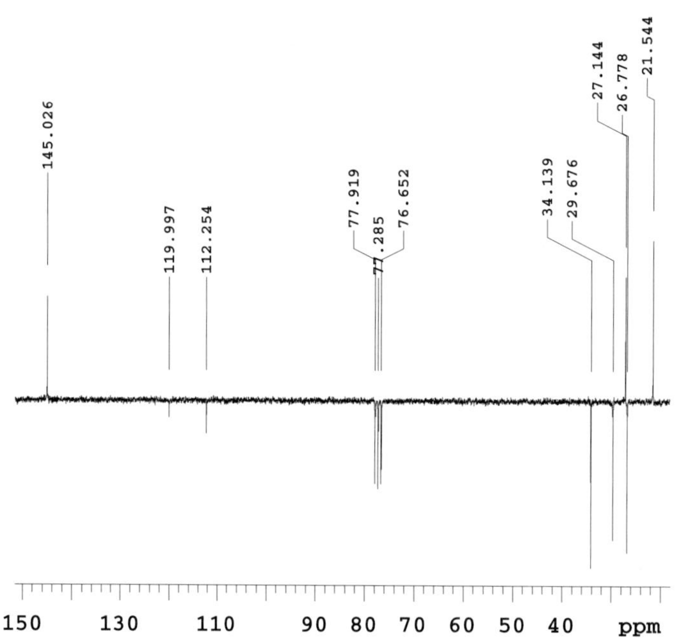

Problem 16: $C_8H_{11}N$
200 MHz, solvent: $CDCl_3$
H,H and C,H correlation

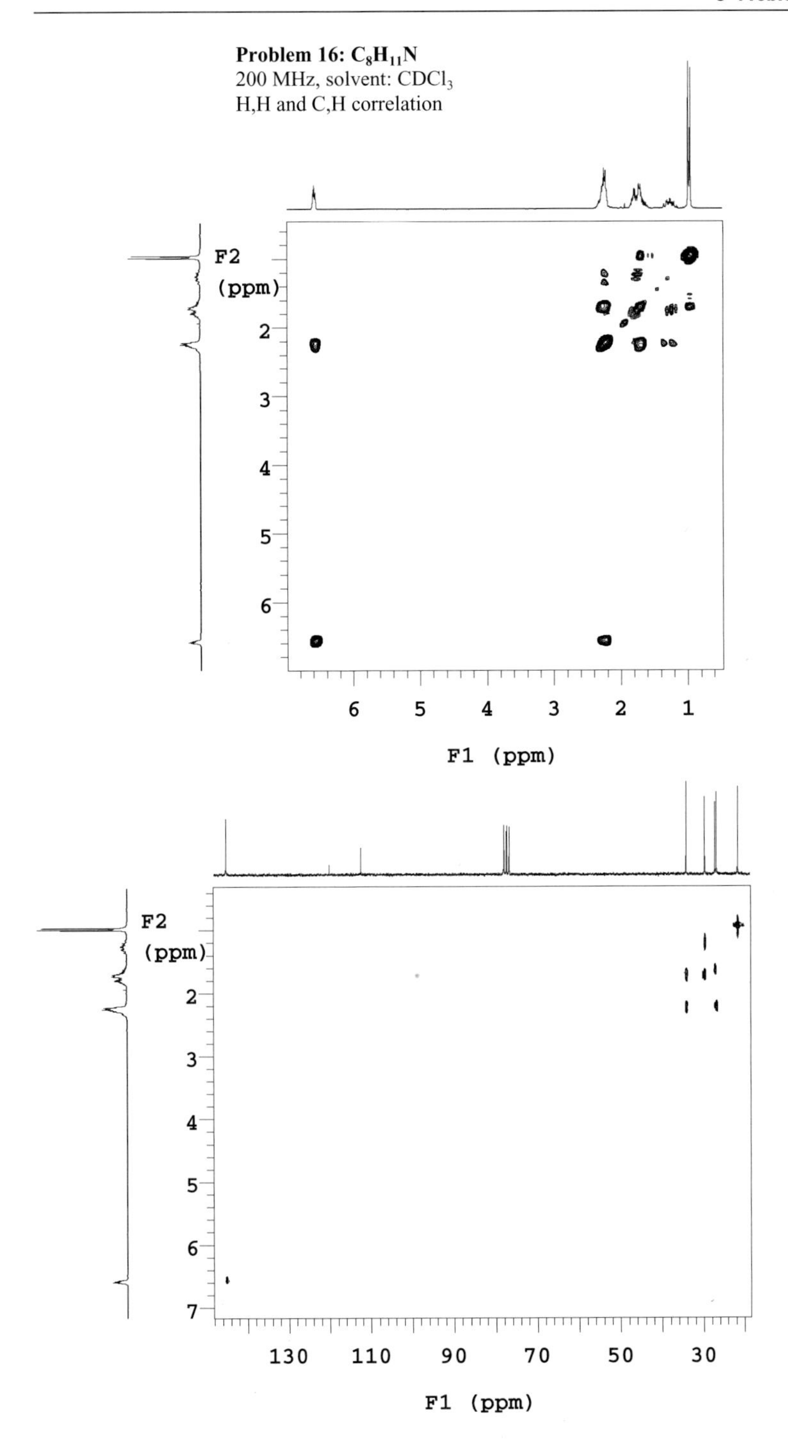

Problem 17: $C_8H_{11}NO_2$
IR: 3379 cm^{-1}
200 MHz, solvent: $CDCl_3$
^1H and APT spectra

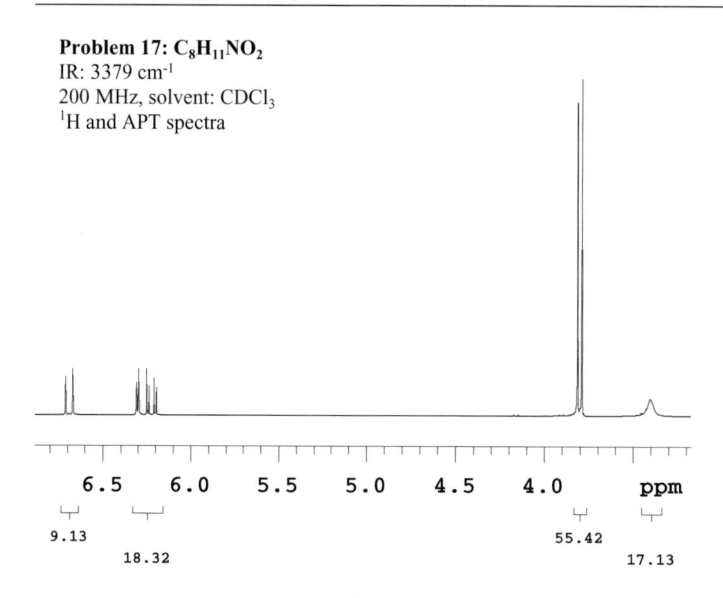

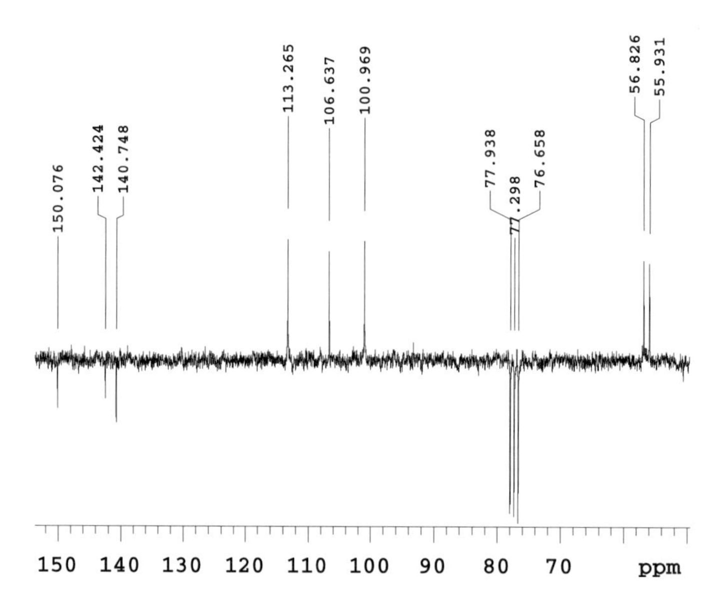

Problem 17: $C_8H_{11}NO_2$
200 MHz, solvent: $CDCl_3$
H,H and C,H correlation

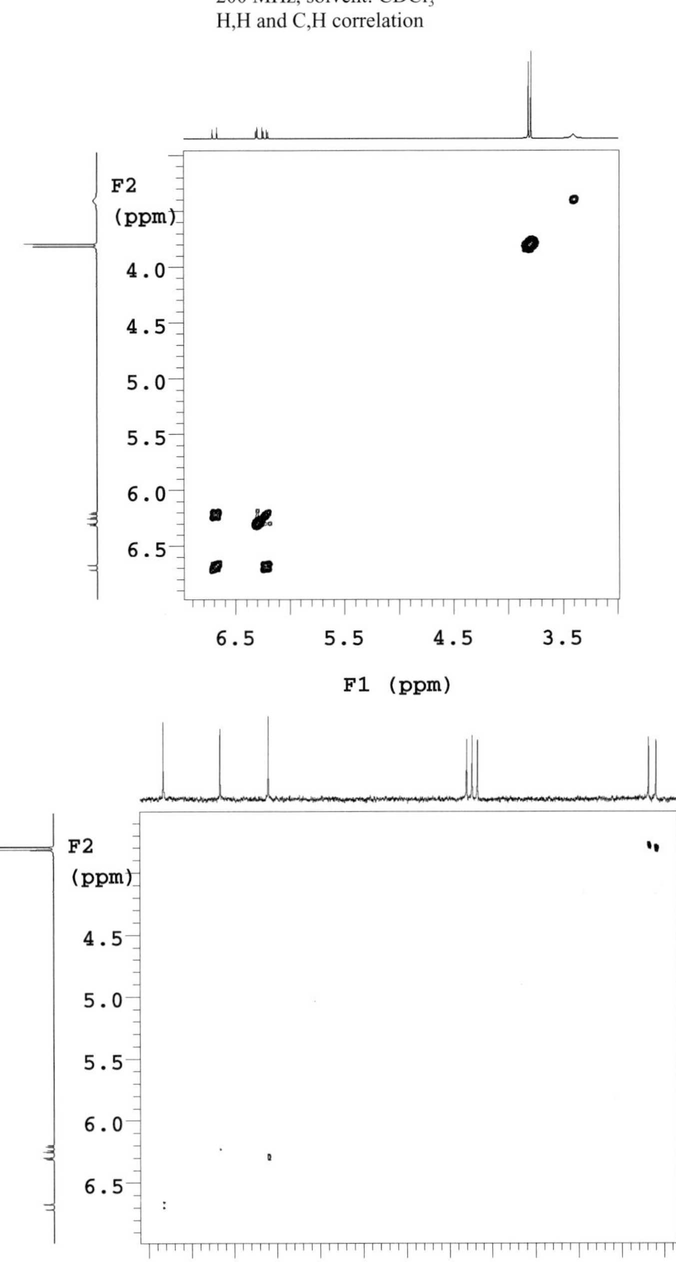

Problem 18: $C_8H_{12}O$
IR: 1680 cm^{-1}
600 MHz, solvent: $CDCl_3$
^1H and ^{13}C spectra

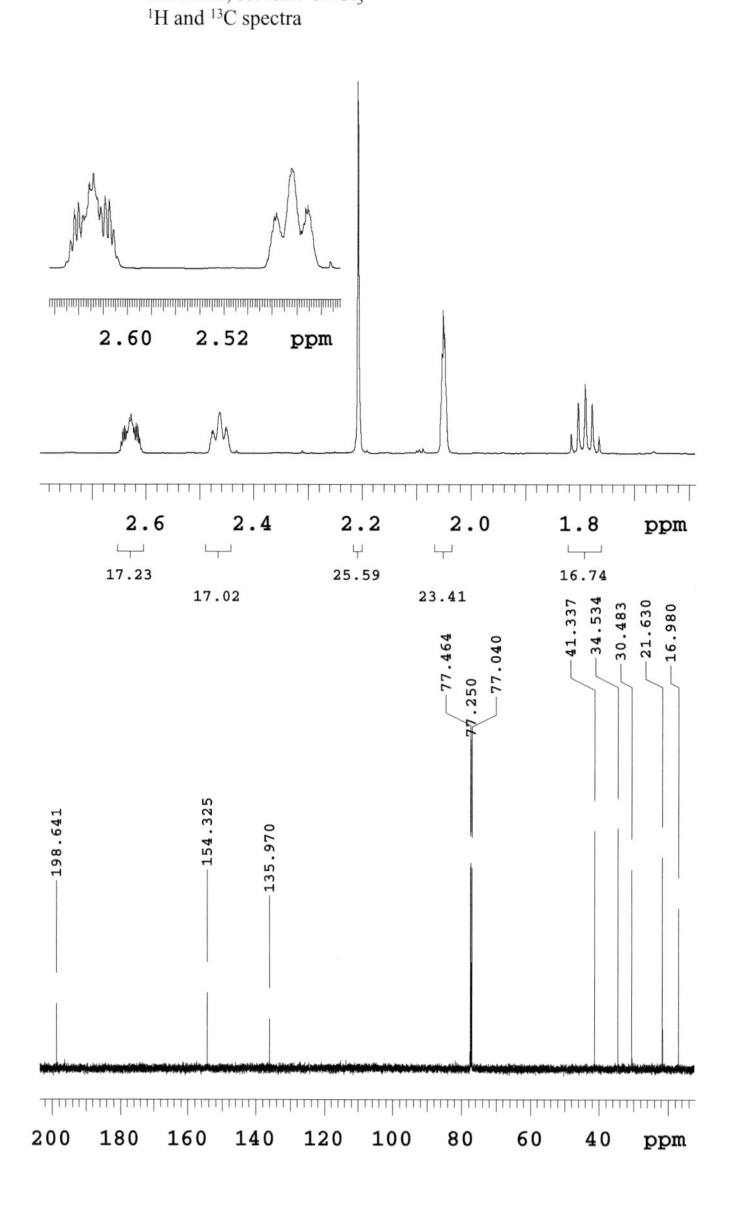

Problem 18: $C_8H_{12}O$
600 MHz, solvent: $CDCl_3$
H,H and C,H correlation

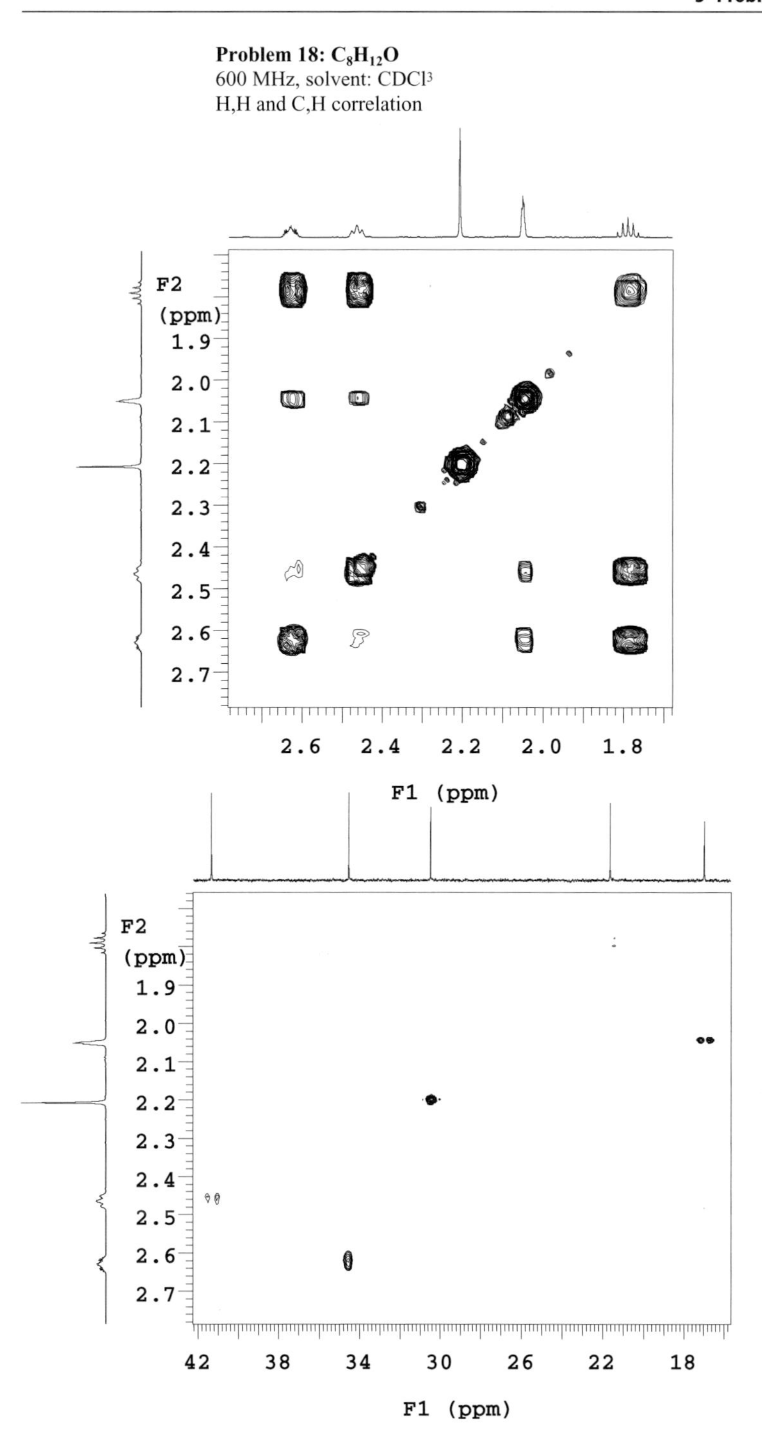

Problem 19: C_9H_8O
IR: 1625, 1681 (strong) cm^{-1}
200 MHz, solvent: $CDCl_3$
^1H and APT spectra

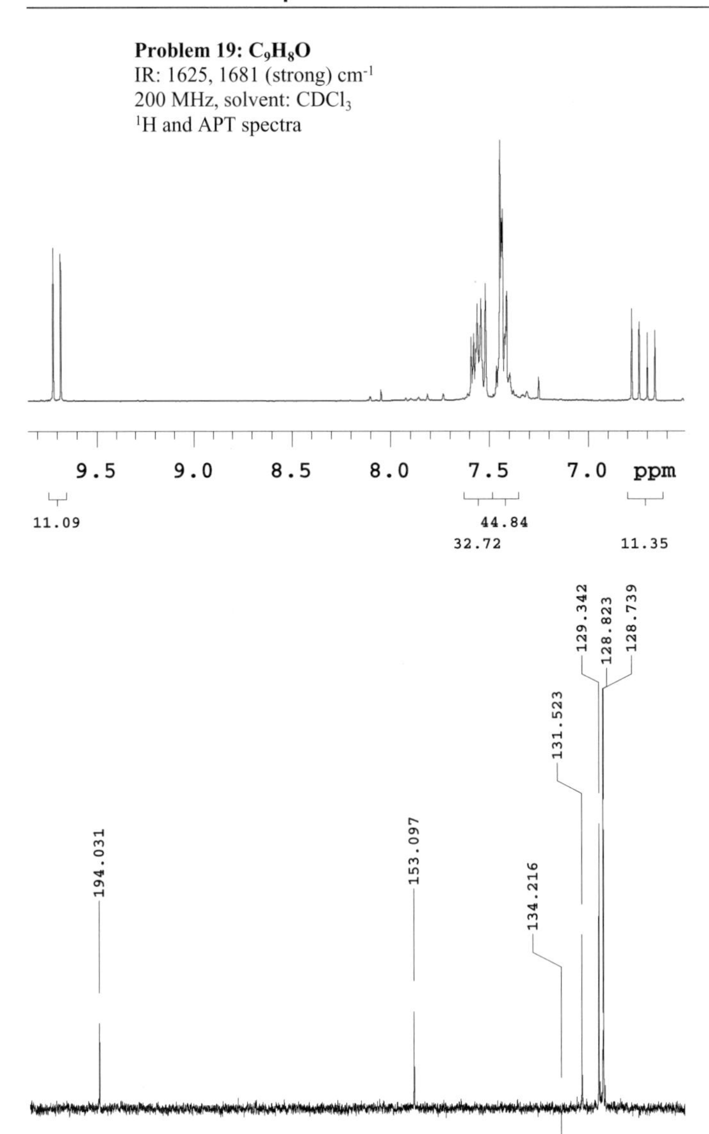

Problem 19: C_9H_8O
200 MHz, solvent: $CDCl_3$
H,H correlation and a part of C,H correlation

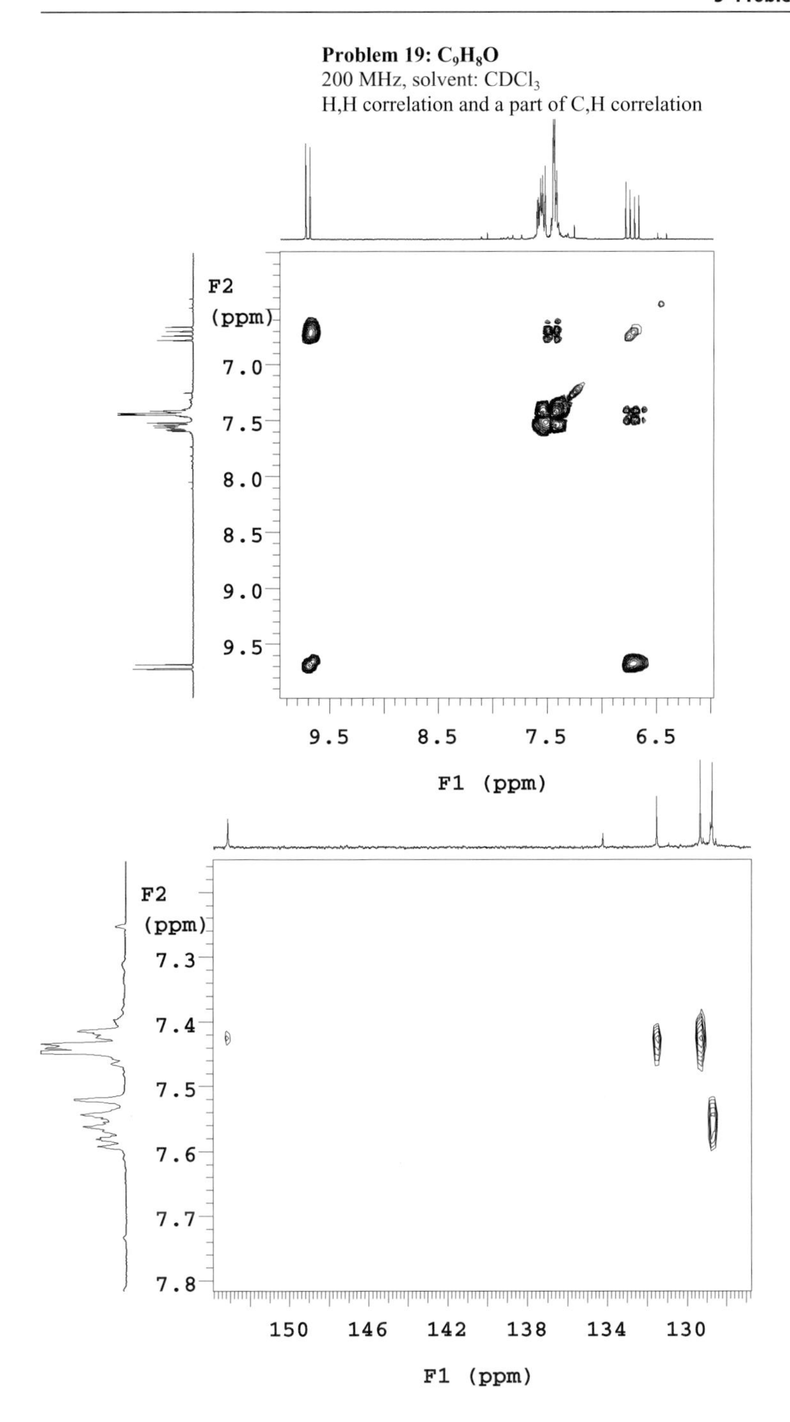

Problem 20: $C_9H_{12}O_3$
IR: 3480 cm^{-1}
400 MHz, solvent: $CDCl_3$
^1H and APT spectra

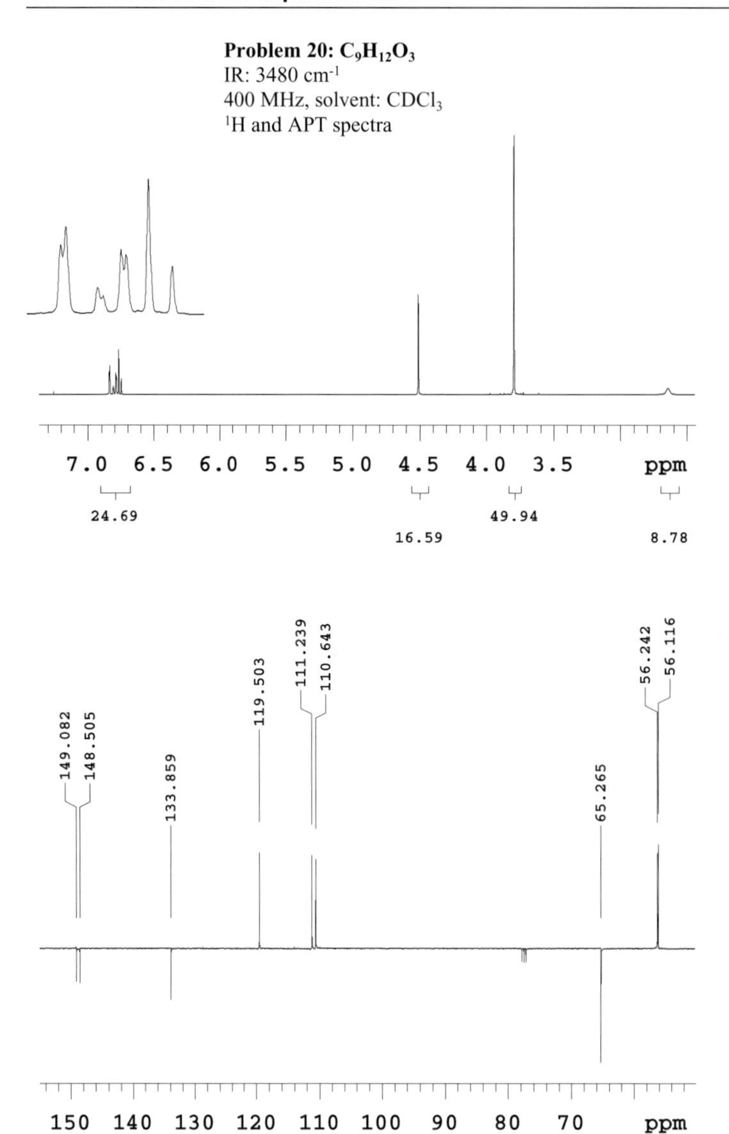

Problem 20: $C_9H_{12}O_3$
400 MHz, solvent: $CDCl_3$
^{13}C spectra and C,H correlation

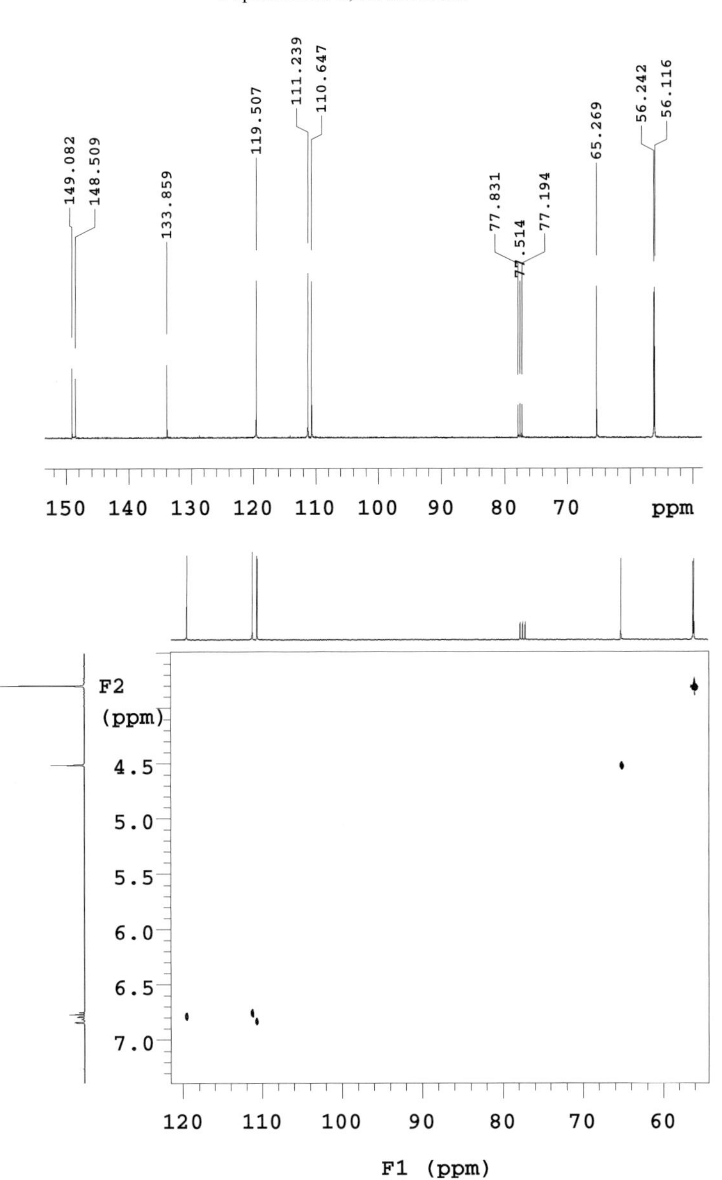

Problem 21: $C_9H_{14}O_2$
IR: 1697, 2900 (broad) cm^{-1}
400 MHz, solvent: $CDCl_3$
^1H and APT spectra

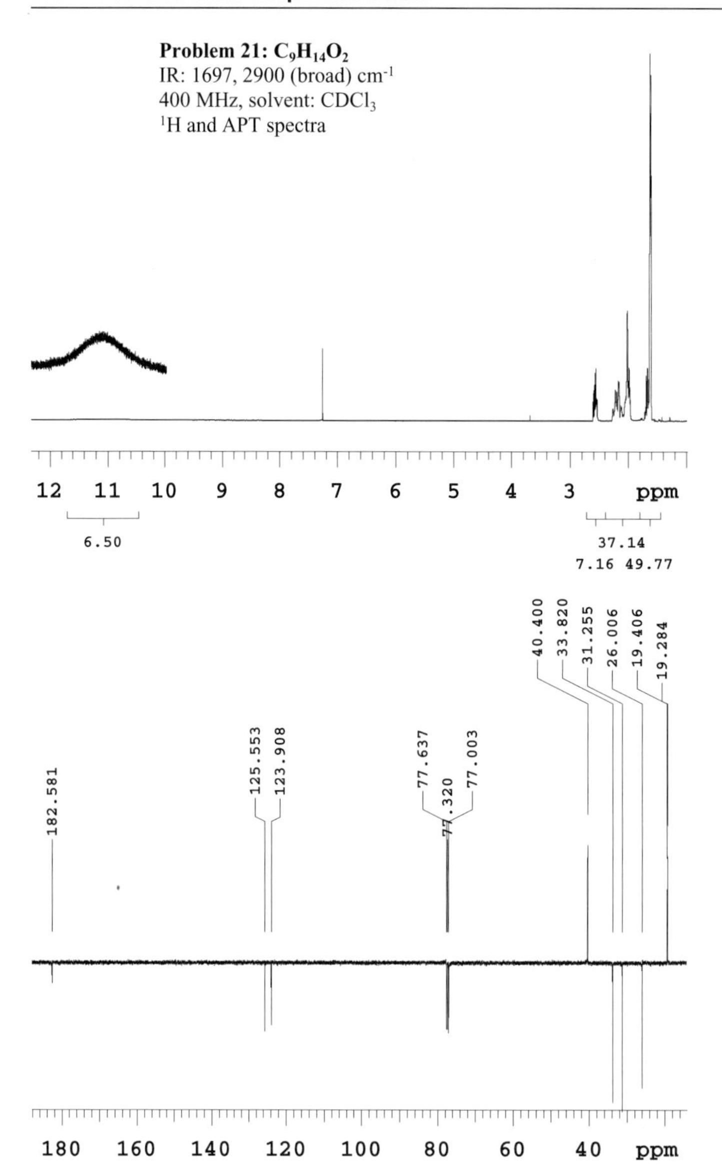

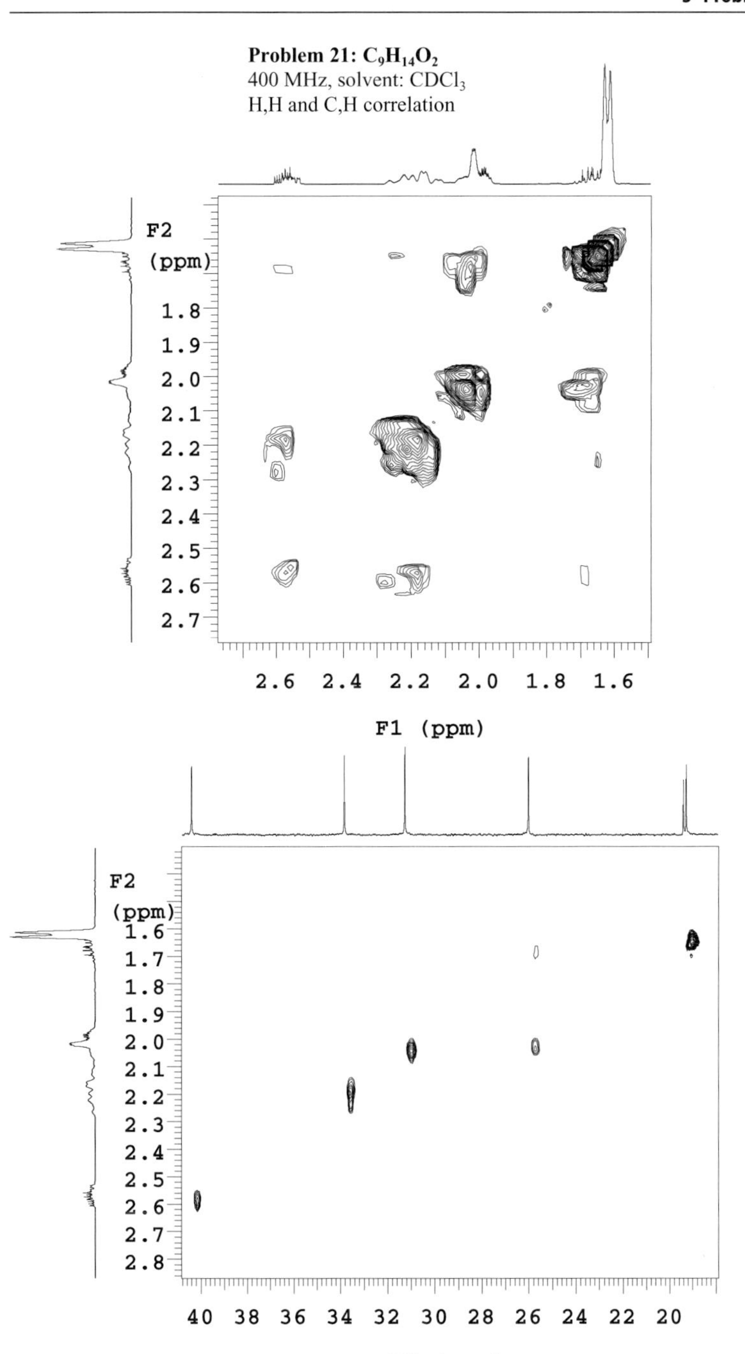
Problem 21: $C_9H_{14}O_2$
400 MHz, solvent: $CDCl_3$
H,H and C,H correlation
F2
(ppm)
1.8
1.9
2.0
2.1
2.2
2.3
2.4
2.5
2.6
2.7
2.6 2.4 2.2 2.0 1.8 1.6
F1 (ppm)
F2
(ppm)
1.6
1.7
1.8
1.9
2.0
2.1
2.2
2.3
2.4
2.5
2.6
2.7
2.8
40 38 36 34 32 30 28 26 24 22 20
F1 (ppm)

Problem 22: $C_{10}H_{14}O$
IR: 1631, 1732 cm^{-1}
200 MHz, solvent: $CDCl_3$
^1H and APT spectra

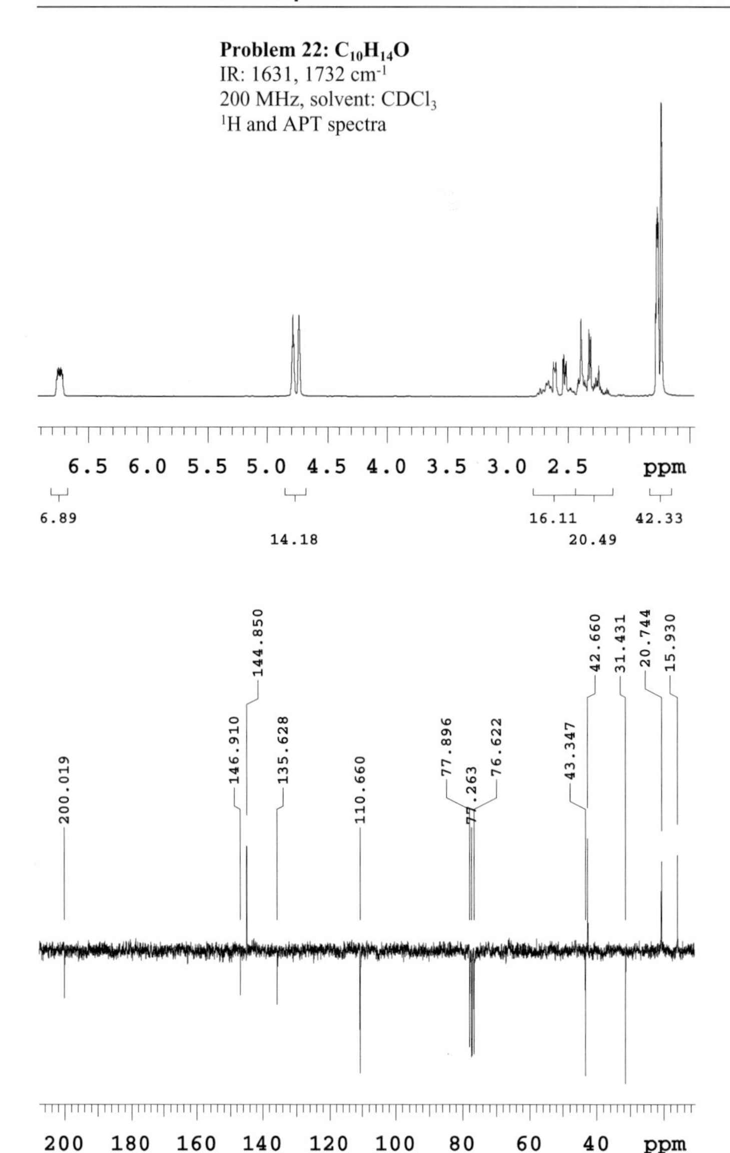

Problem 22: $C_{10}H_{14}O$
200 MHz, solvent: $CDCl_3$
H,H and C,H correlation

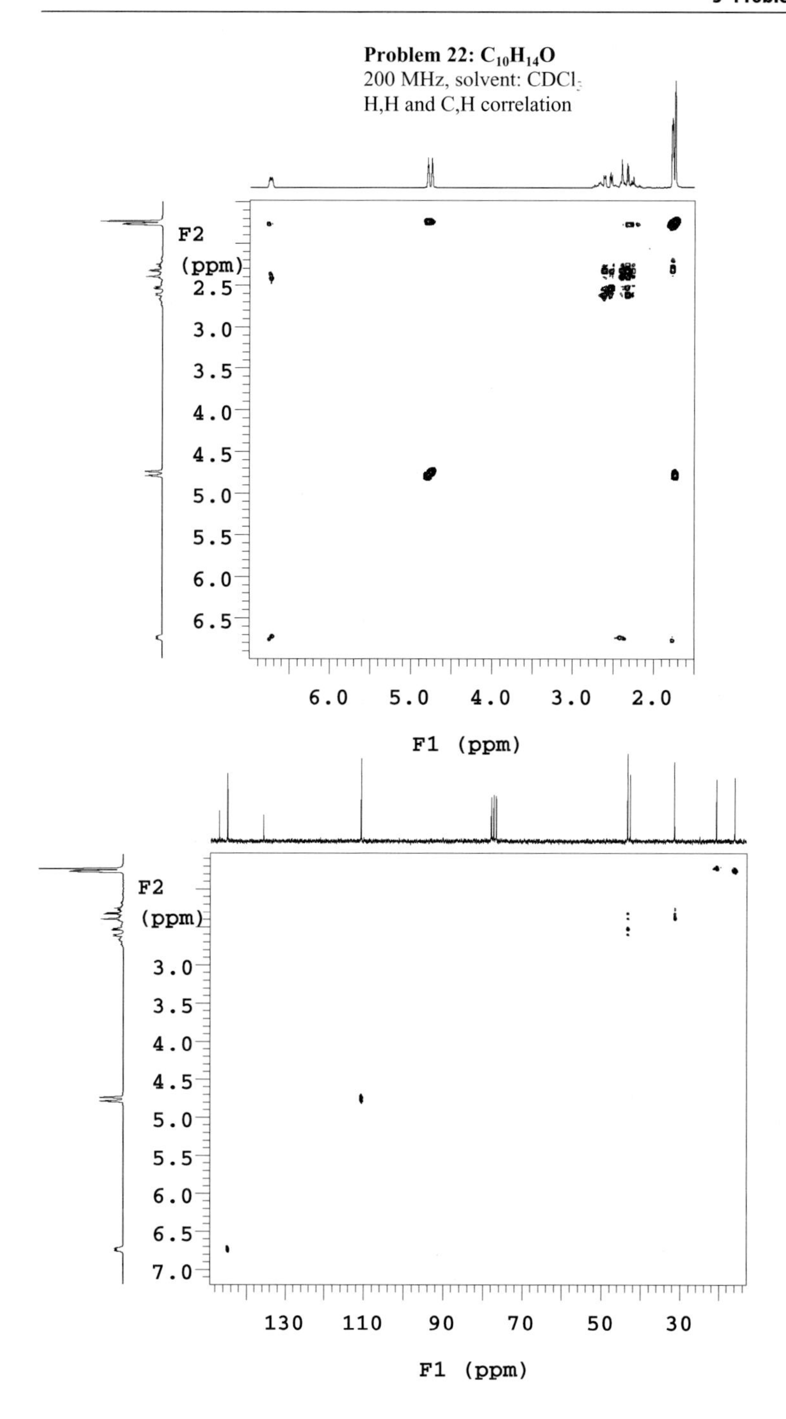

Problem 23: $C_{10}H_{16}O_2$
IR: 1725 cm^{-1}
500 MHz, solvent: $CDCl_3$
^1H spectrum and C,H correlation

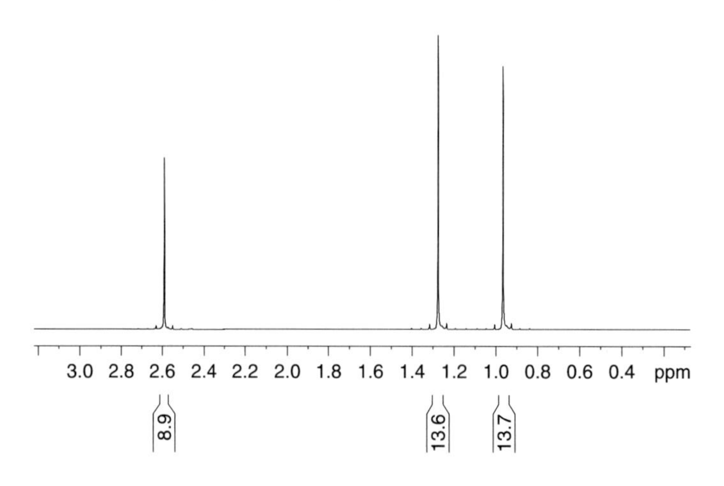

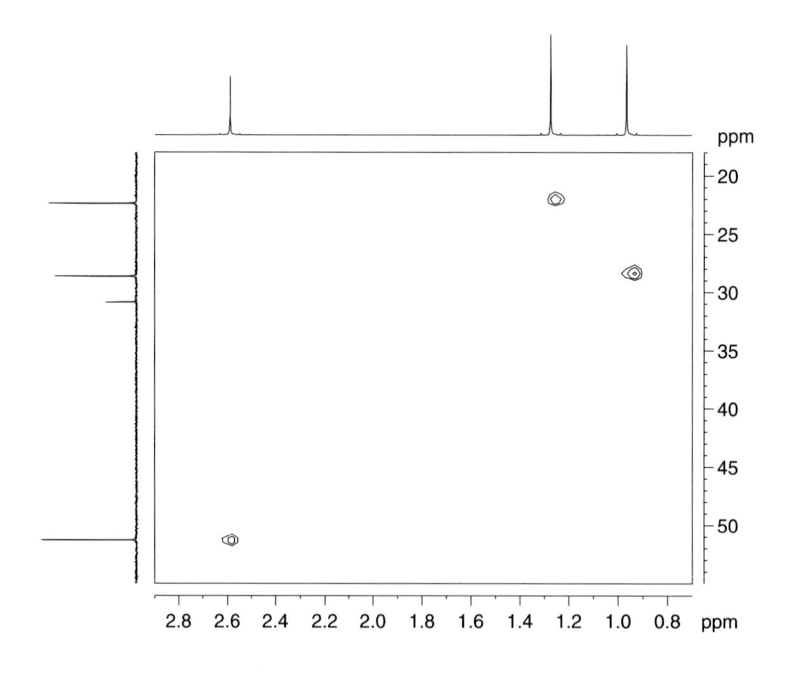

Problem 23: $C_{10}H_{16}O_2$
500 MHz, solvent: $CDCl_3$
^{13}C and APT spectra

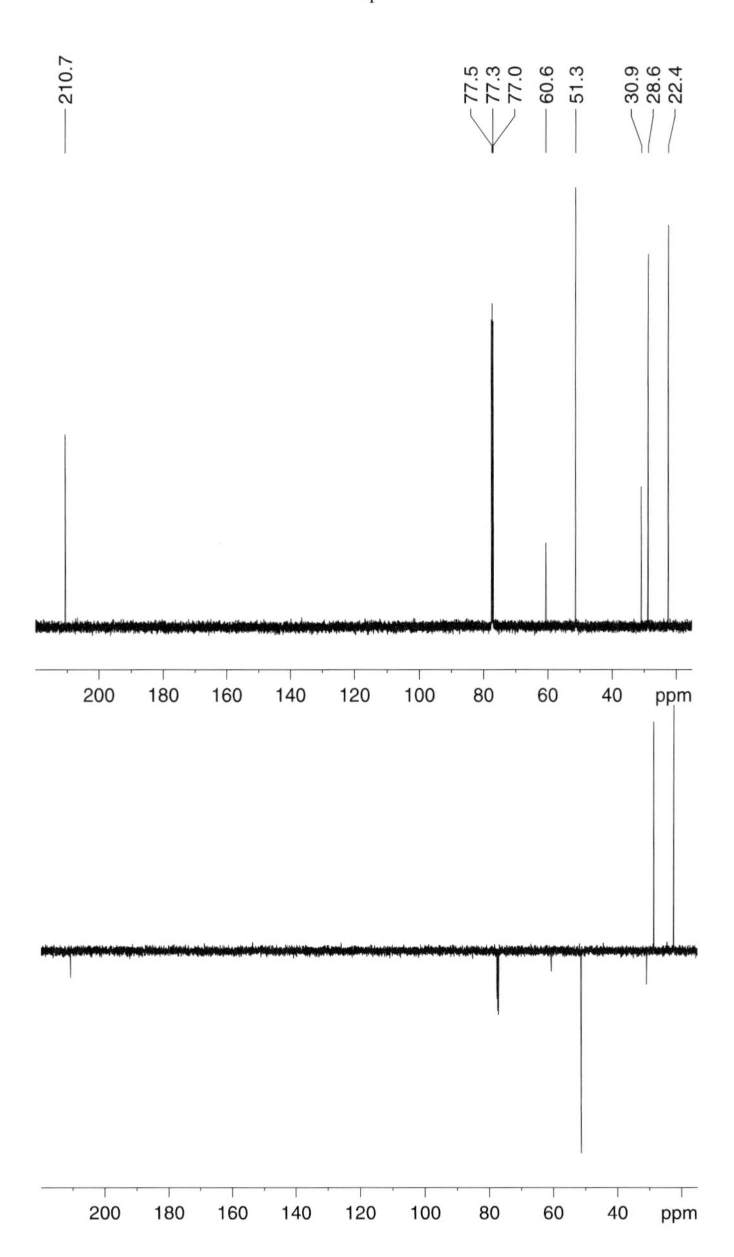

Problem 24: $C_{10}H_{18}O_6$
IR: 1730 (strong), 3340 (broad) cm^{-1}
500 MHz, solvent: $CDCl_3$
^1H and APT spectra

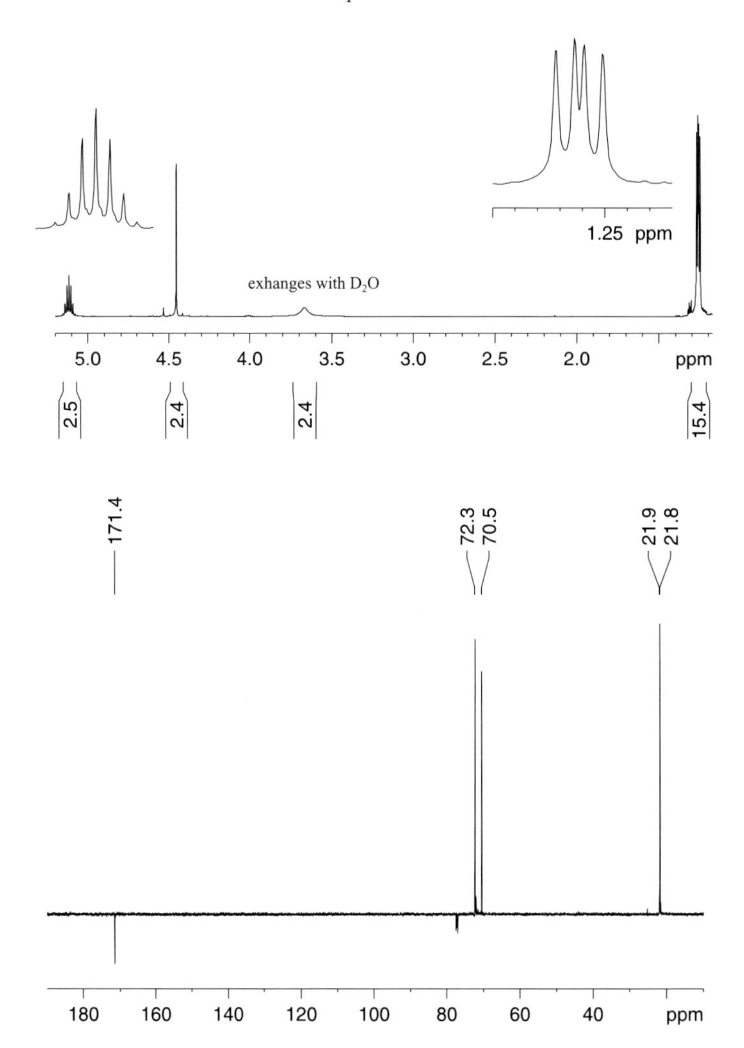

Problem 24: $C_{10}H_{18}O_6$
500 MHz, solvent: $CDCl_3$
H,H and C,H correlation

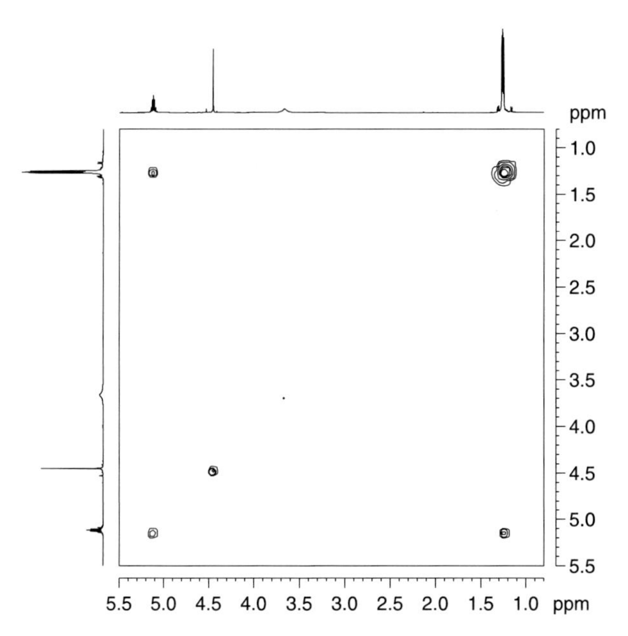

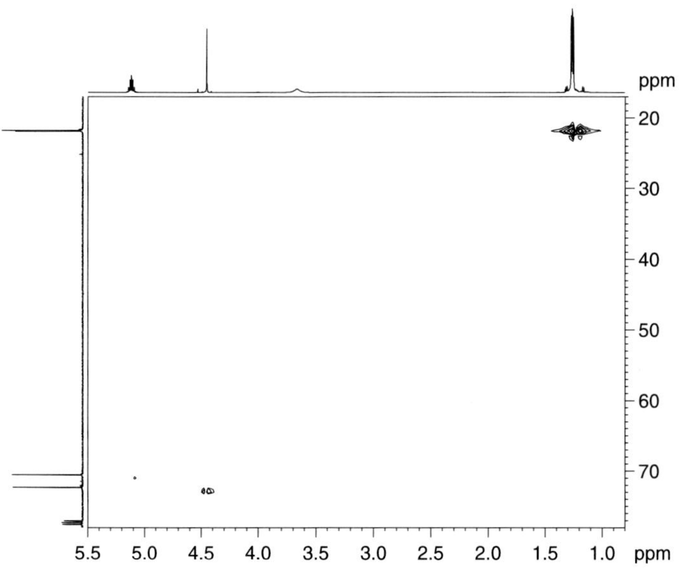

Problem 25: $C_{11}H_8O_2$
IR: 1670 (strong), 3360 cm^{-1}
500 MHz, solvent: $CDCl_3$
^1H and APT spectra

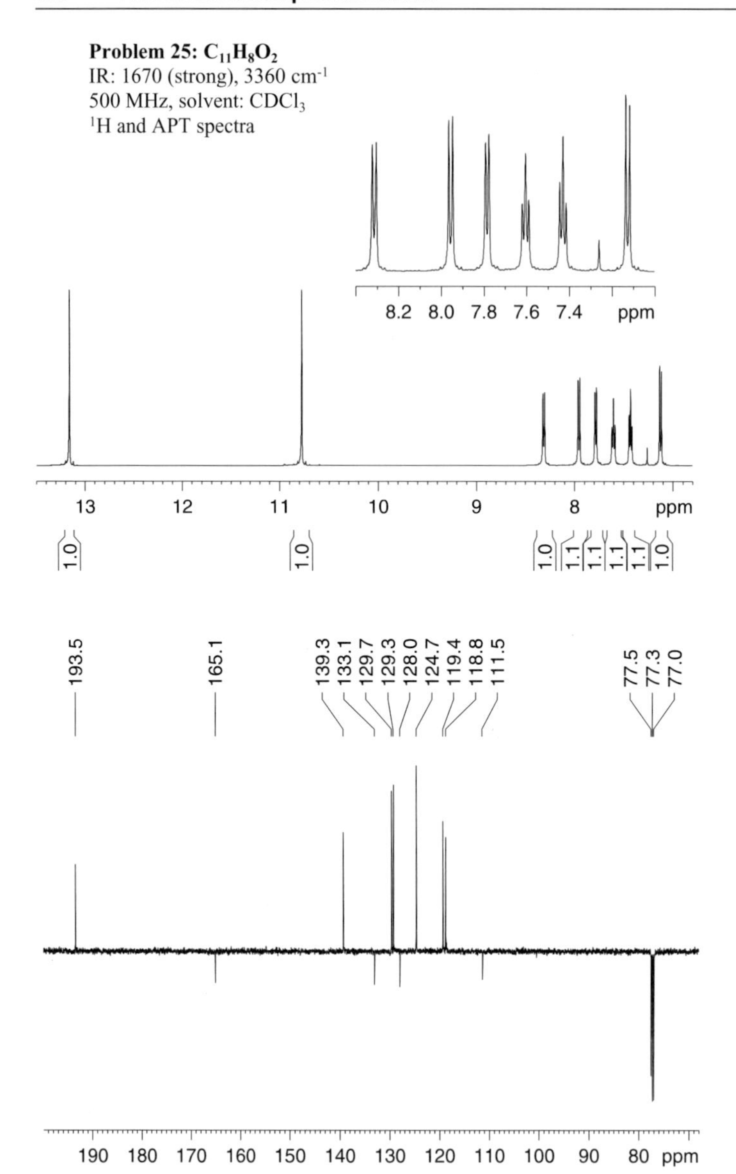

Problem 25: $C_{11}H_8O_2$
500 MHz, solvent: $CDCl_3$
H,H and C,H correlation

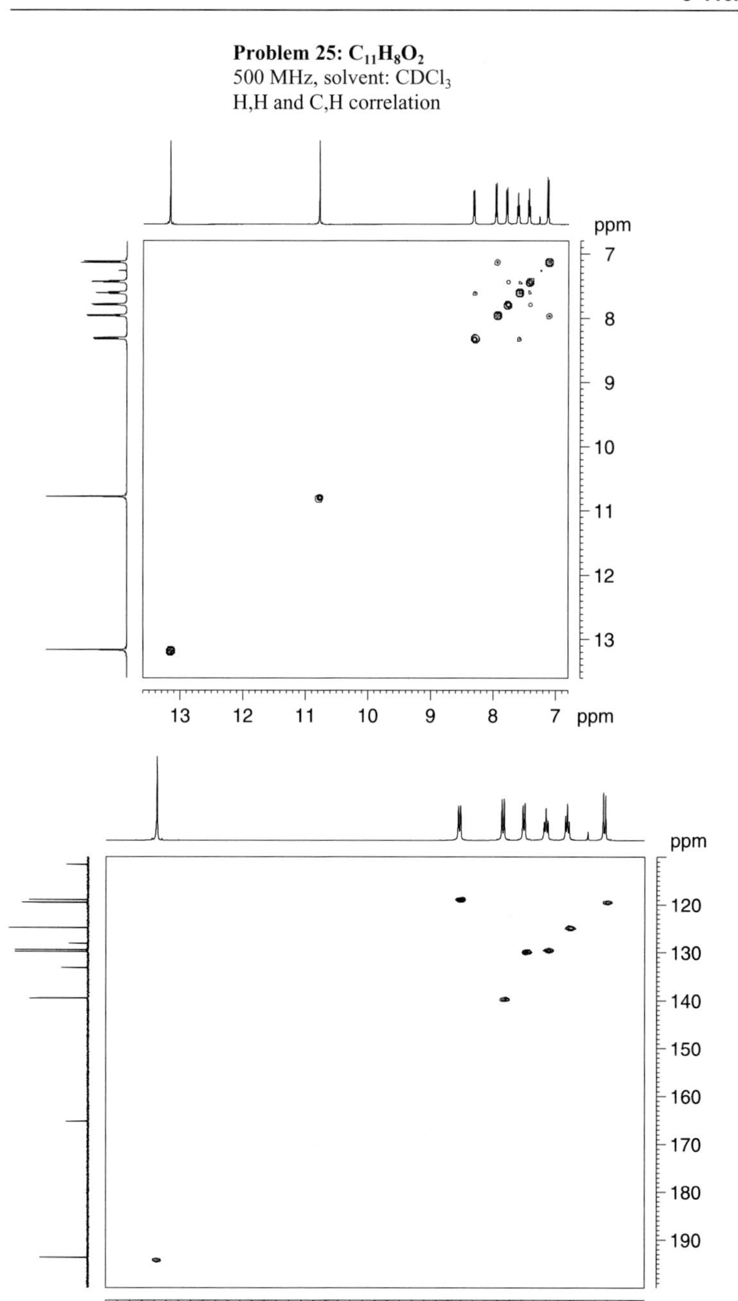

Problem 26: $C_{11}H_{18}O_4$
IR: 1732 cm^{-1}
200 MHz, solvent: $CDCl_3$
^1H with expansion and APT spectra

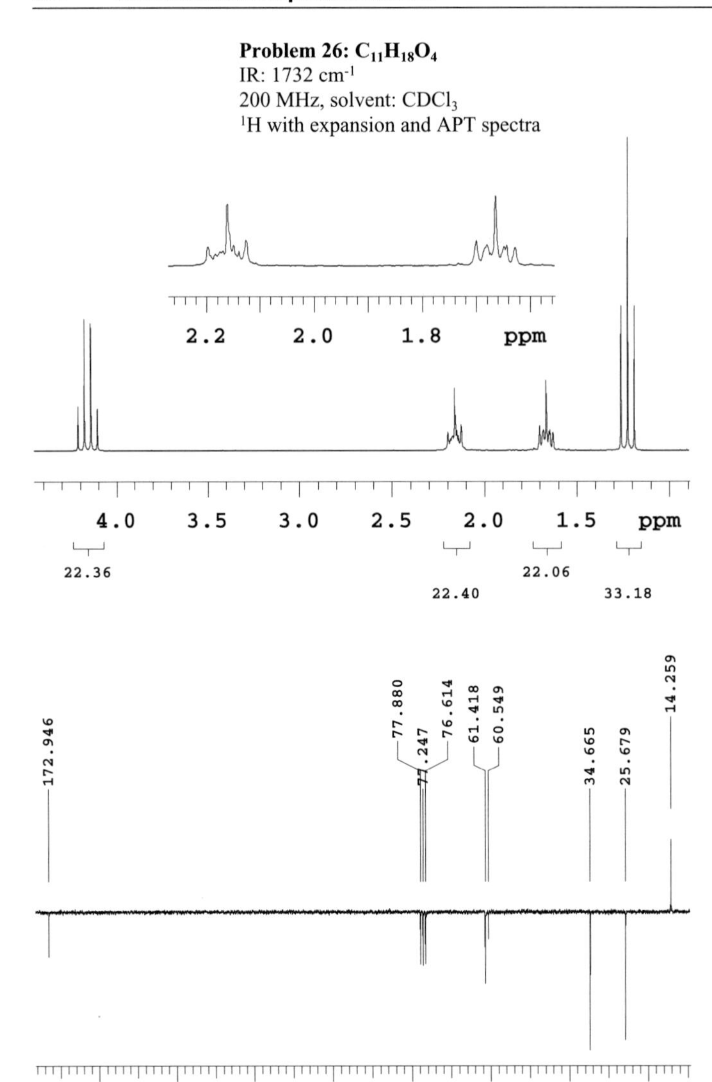

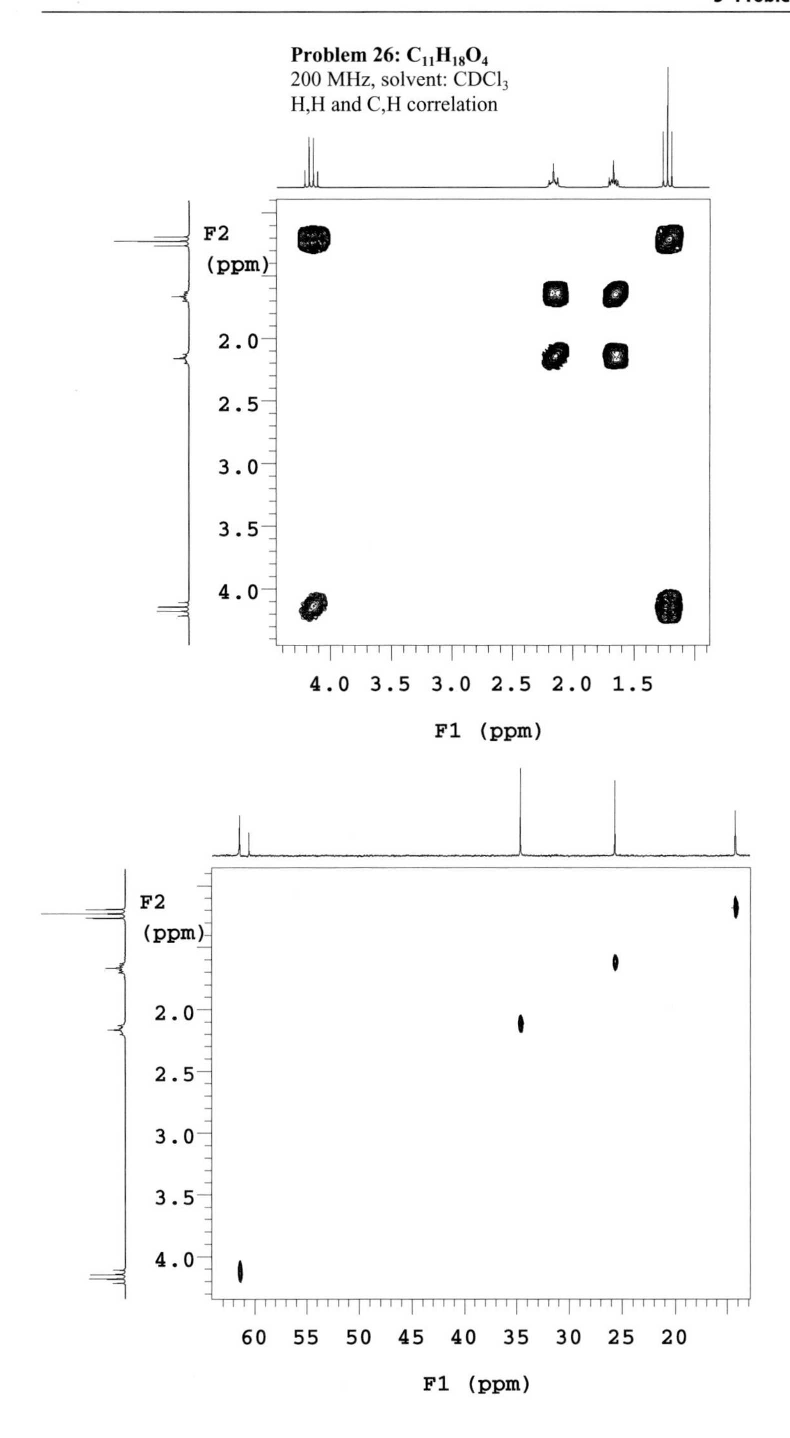
Problem 26: $C_{11}H_{18}O_4$
200 MHz, solvent: $CDCl_3$
H,H and C,H correlation
F2
(ppm)
2.0
2.5
3.0
3.5
4.0
4.0 3.5 3.0 2.5 2.0 1.5
F1 (ppm)
F2
(ppm)
2.0
2.5
3.0
3.5
4.0
60 55 50 45 40 35 30 25 20
F1 (ppm)

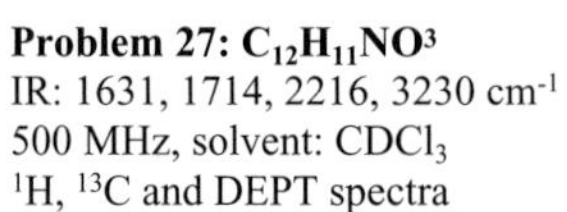

Problem 27: $C_{12}H_{11}NO^{3}$
IR: 1631, 1714, 2216, 3230 cm^{-1}
500 MHz, solvent: $CDCl_3$
^{1}H, ^{13}C and DEPT spectra

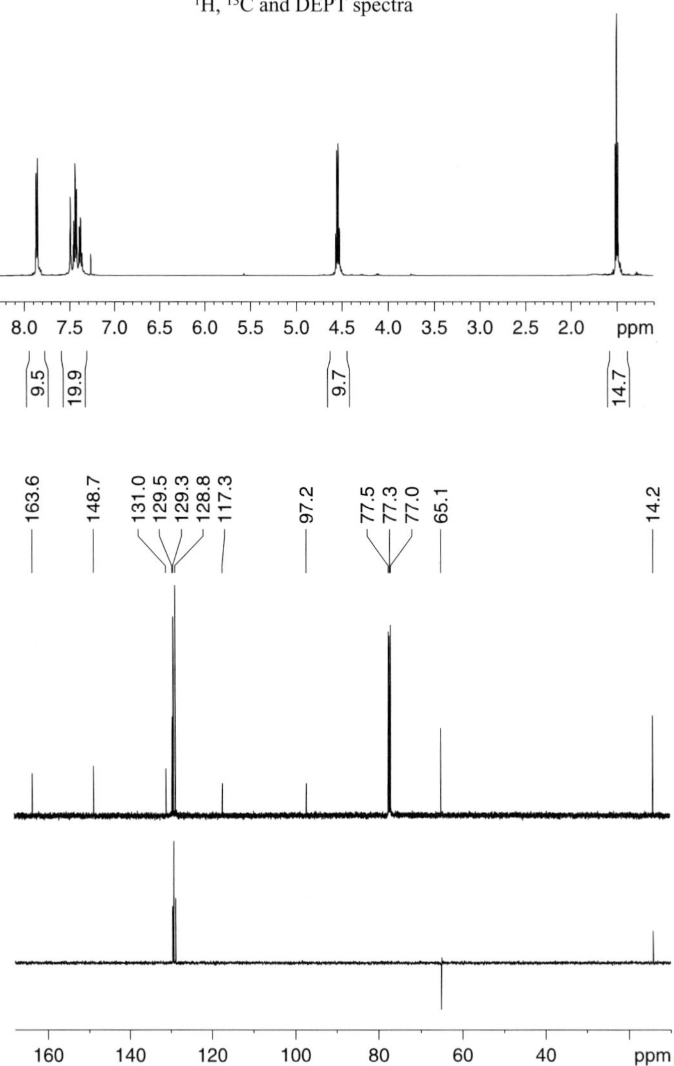

Problem 27: $C_{12}H_{11}NO_3$
500 MHz, solvent: $CDCl_3$
H,H and C,H correlation

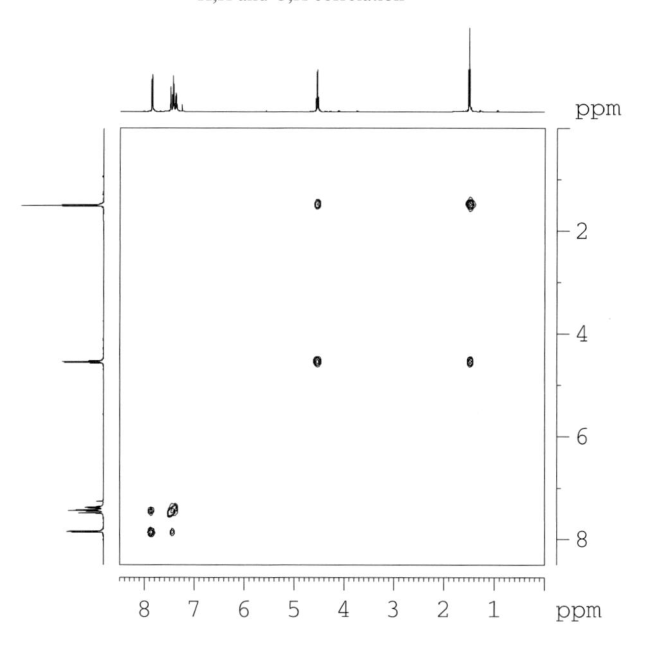

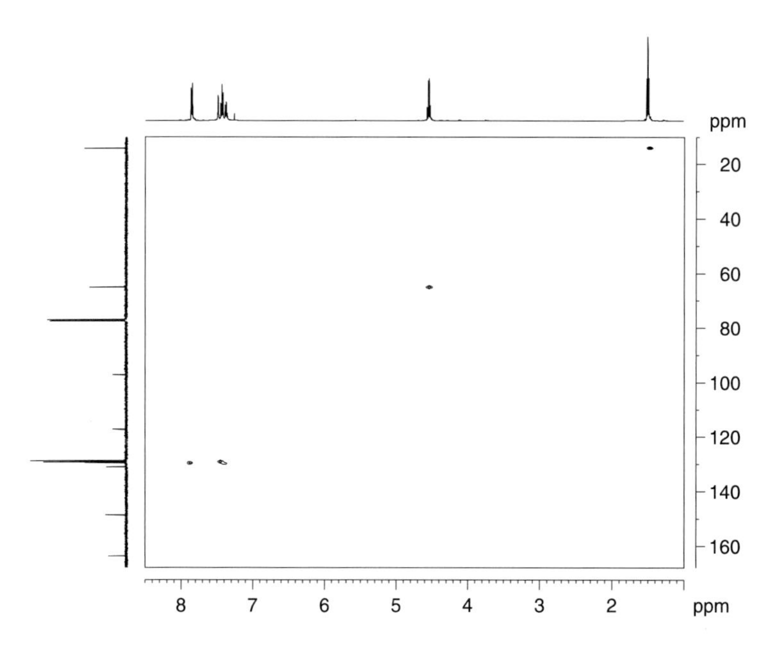

Problem 28: $C_{12}H_{20}O_2$
IR: 1680, 1740 cm^{-1}
500 MHz, solvent: $CDCl_3$
^1H and APT spectra

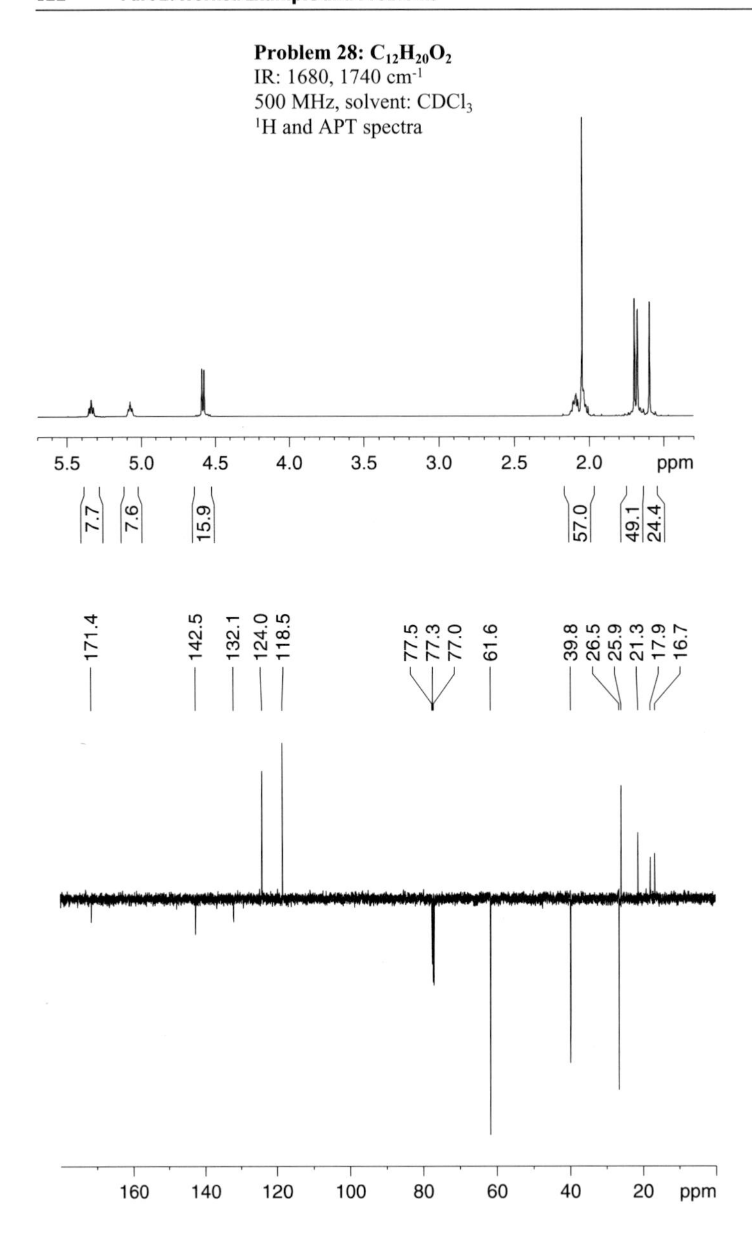

Problem 28: $C_{12}H_{20}O_2$
500 MHz, solvent: $CDCl_3$
H,H and C,H correlation

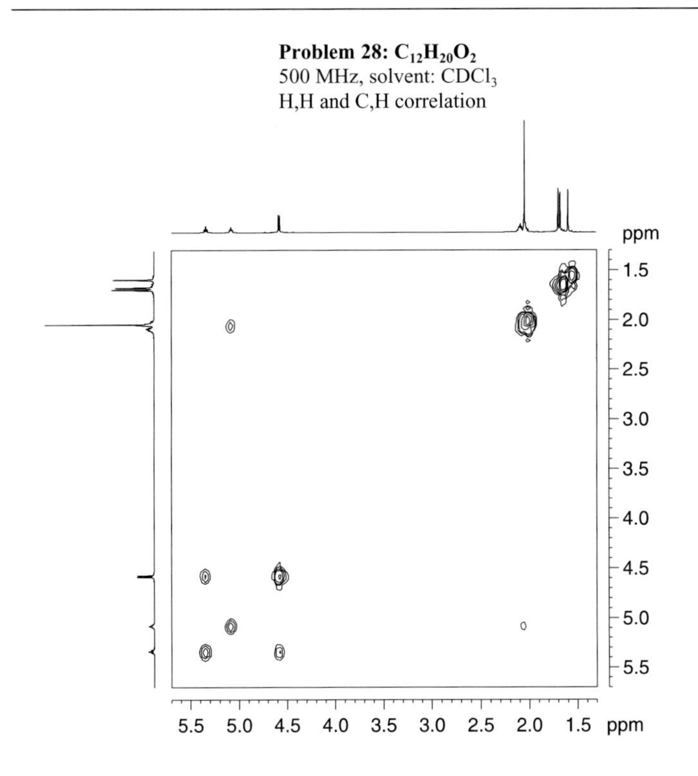

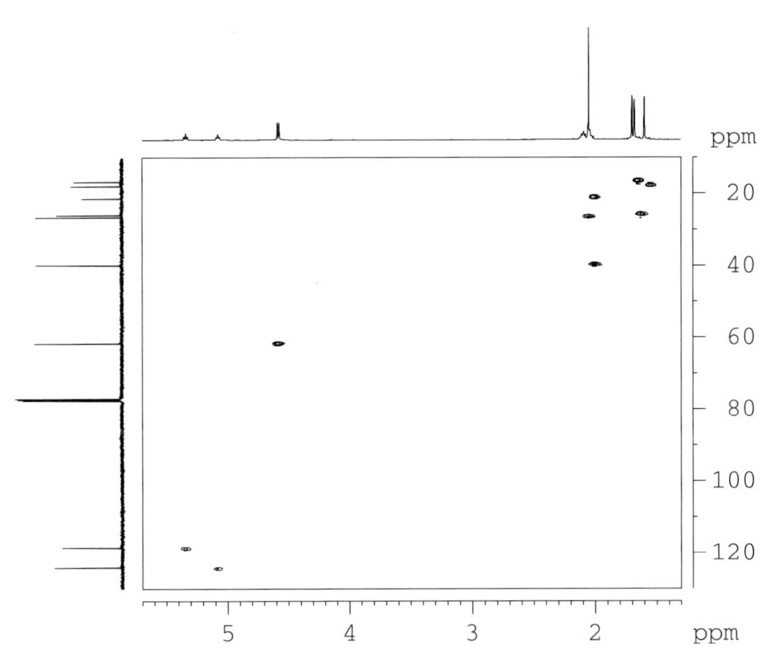

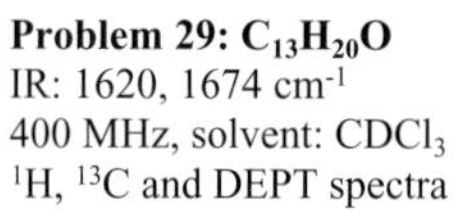

Problem 29: $C_{13}H_{20}O$

IR: 1620, 1674 cm^{-1}

400 MHz, solvent: $CDCl_3$

^1H, ^{13}C and DEPT spectra

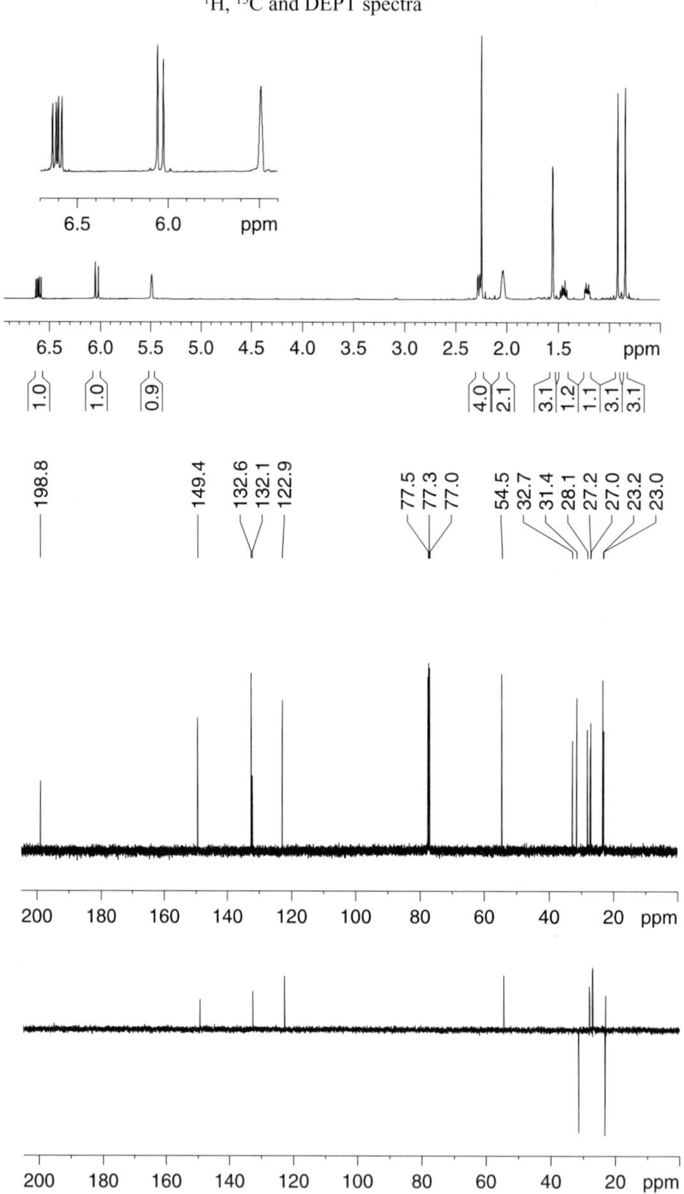

Problem 29: $C_{13}H_{20}O$
400 MHz, solvent: $CDCl_3$
H,H and C,H correlation

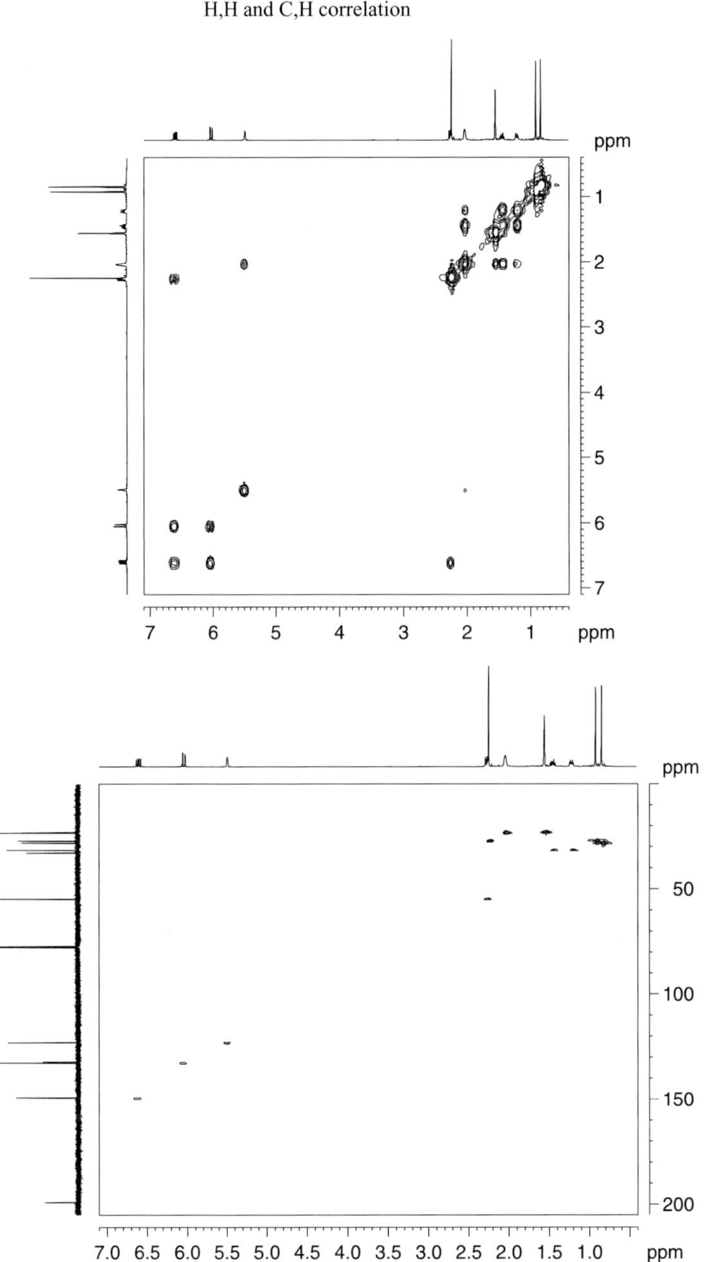

Problem 30: $C_{14}H_{18}N_2$
IR: no bands characteristic of functional groups
500 MHz, solvent: $CDCl_3$
^{1}H, ^{13}C and APT spectra

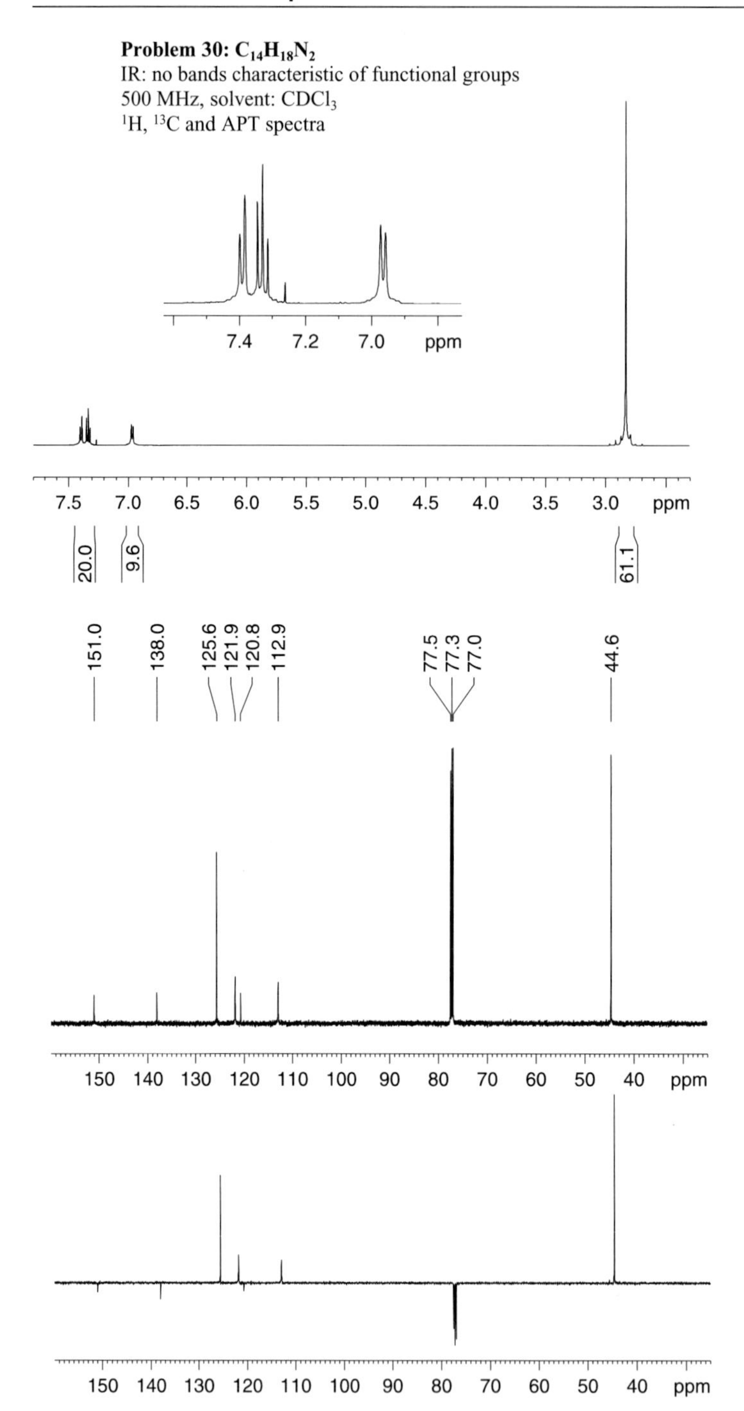

Problem 30: $C_{14}H_{18}N_2$
500 MHz, solvent: $CDCl_3$
H,H and C,H correlation

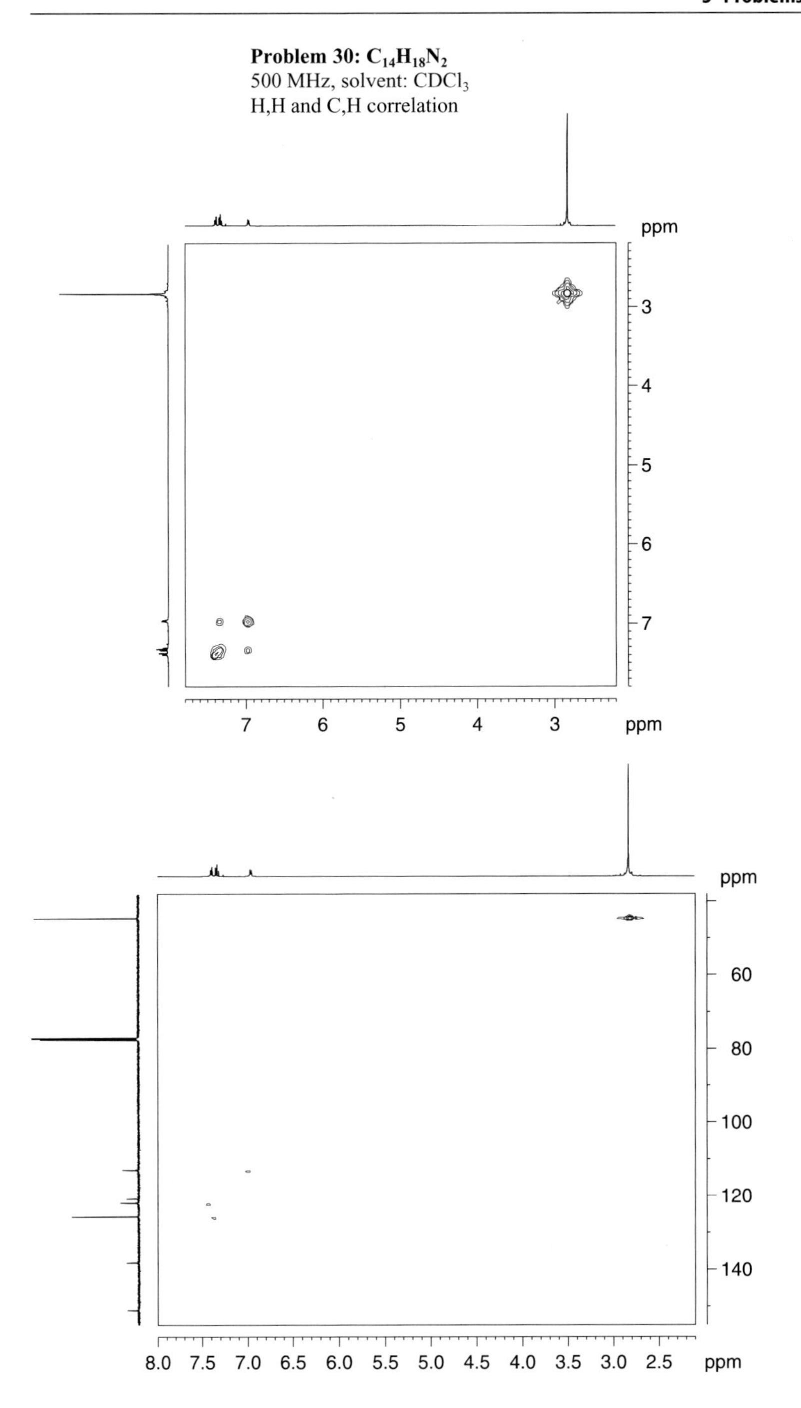

Problem 31: $C_{14}H_{18}O_4$
IR: 1735 cm^{-1}
200 MHz, solvent: $CDCl_3$
^1H with expansion and APT spectra

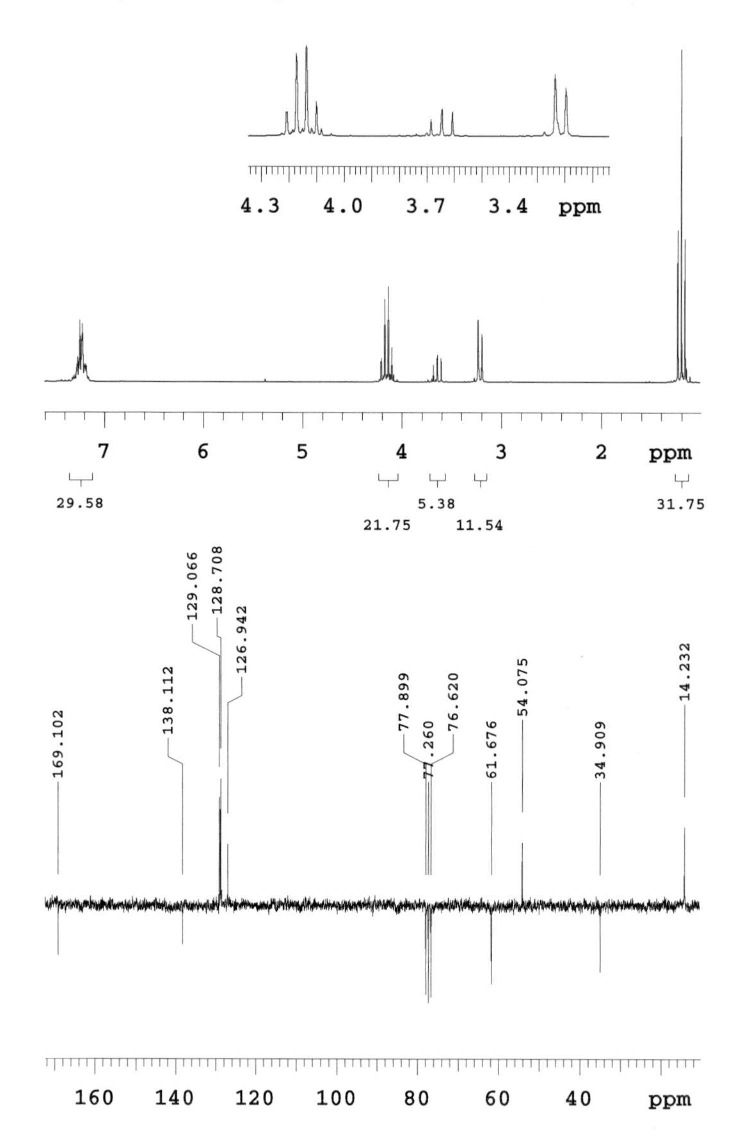

Problem 31: $C_{14}H_{18}O_4$
200 MHz, solvent: $CDCl_3$
H,H and C,H correlation

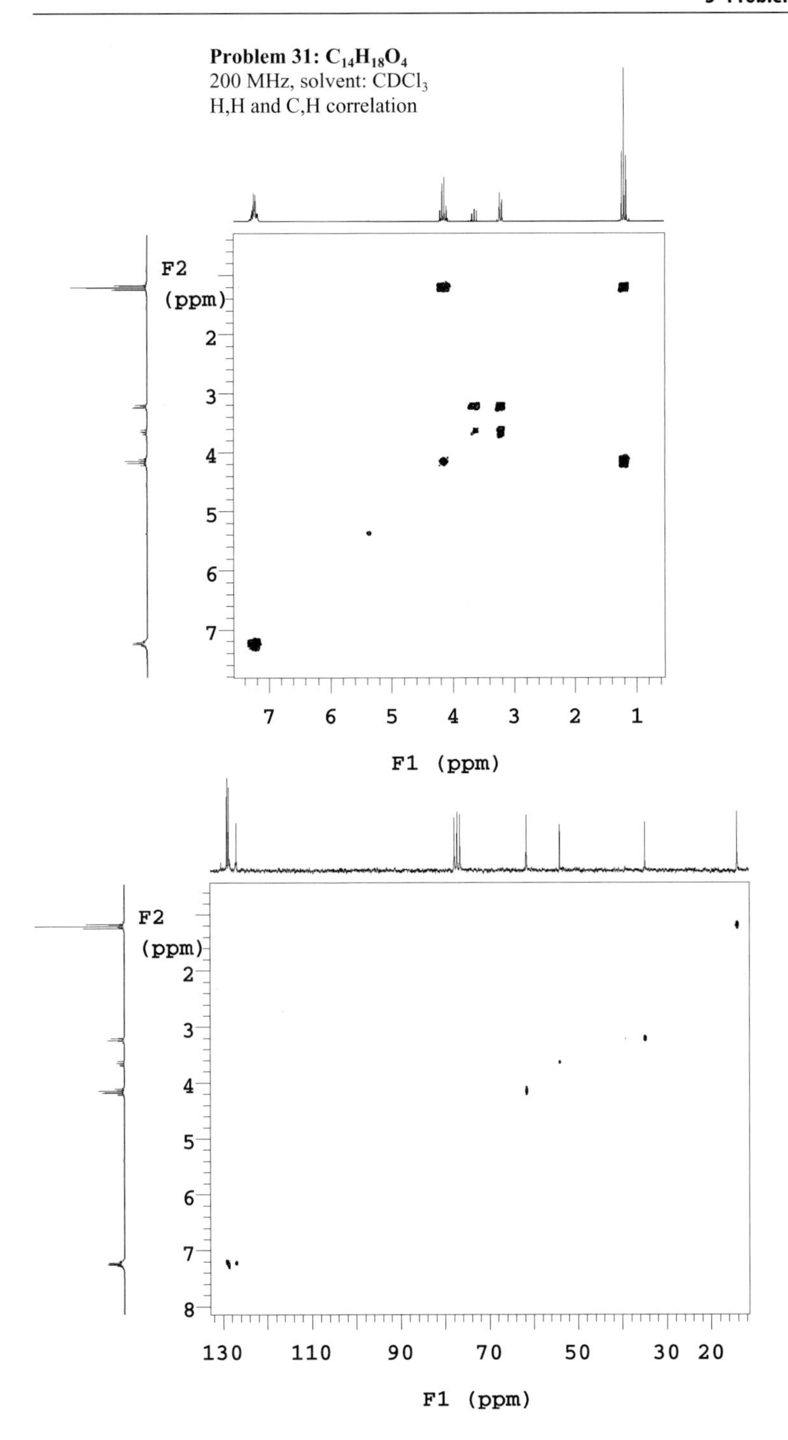

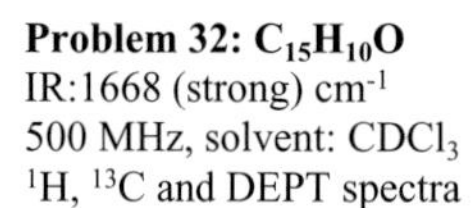

Problem 32: $C_{15}H_{10}O$
IR:1668 (strong) cm^{-1}
500 MHz, solvent: $CDCl_3$
^1H, ^{13}C and DEPT spectra

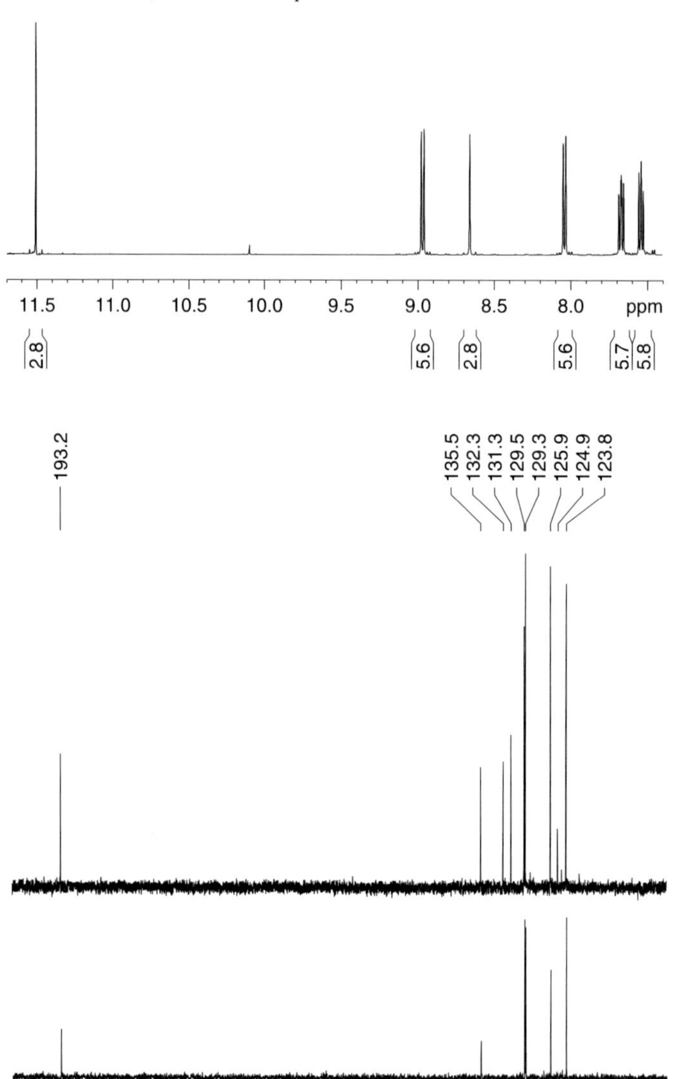

Problem 32: $C_{15}H_{10}O$
500 MHz, solvent: $CDCl_3$
H,H correlation
aromatic part of the C,H correlation

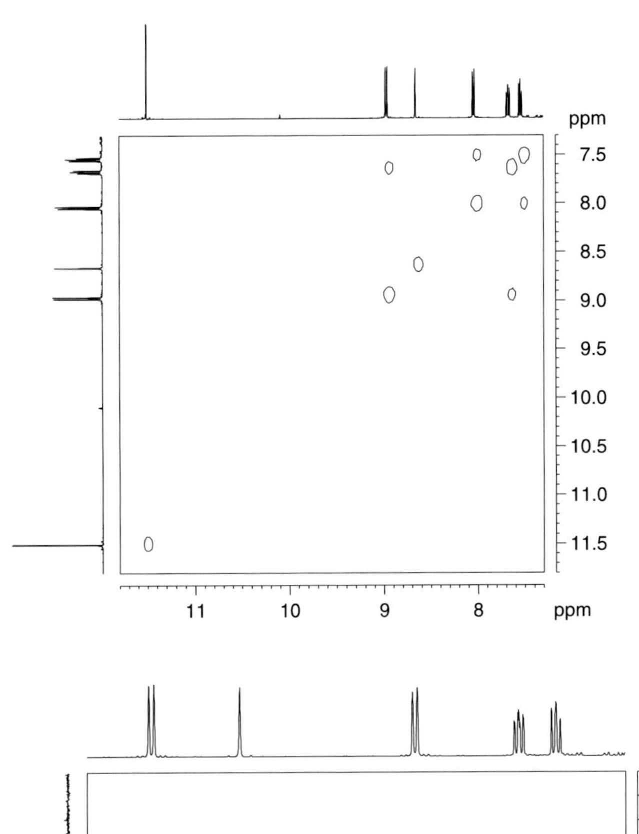

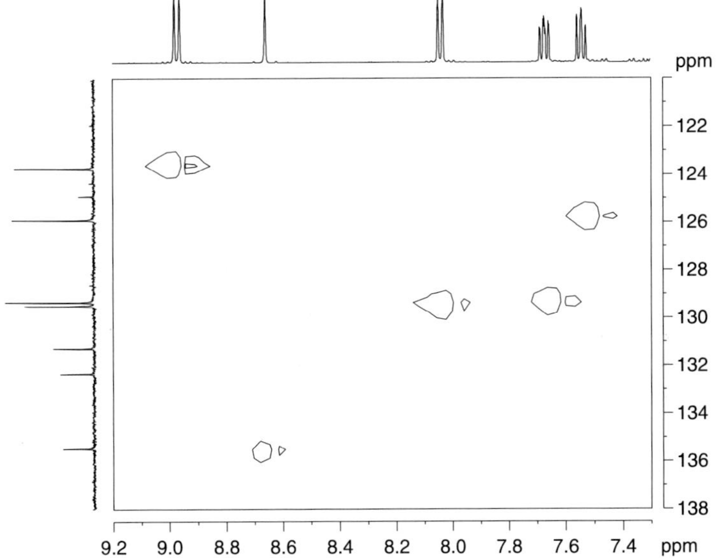

Problem 33: $C_{15}H_{26}O$
IR: 1668, 3366 (broad) cm^{-1}
500 MHz, solvent: $CDCl_3$
^1H and APT spectra

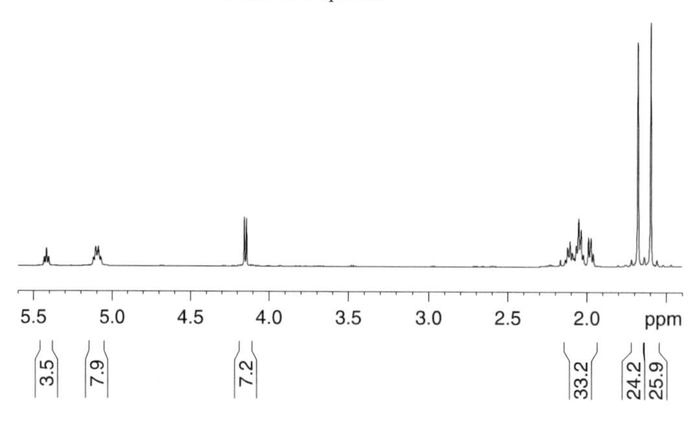

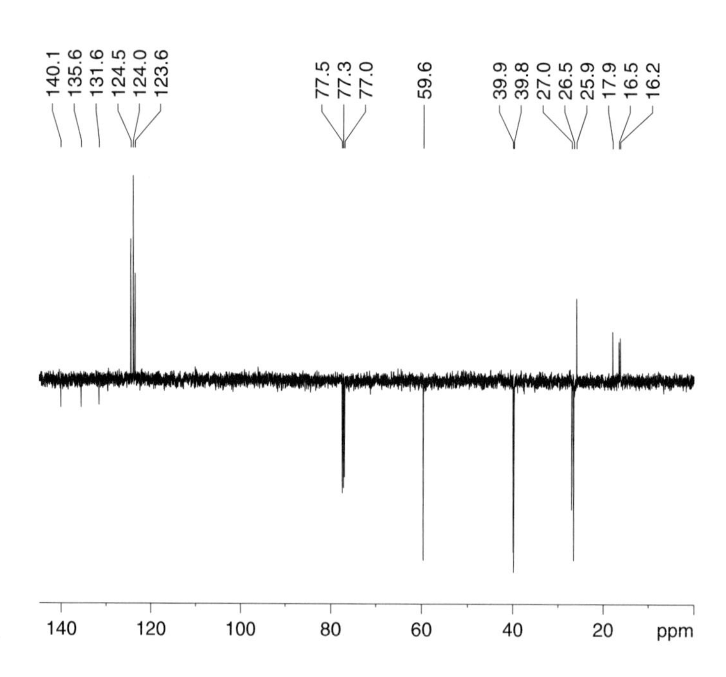

Problem 33: $C_{15}H_{26}O$
500 MHz, solvent: $CDCl_3$
H,H and C,H correlation

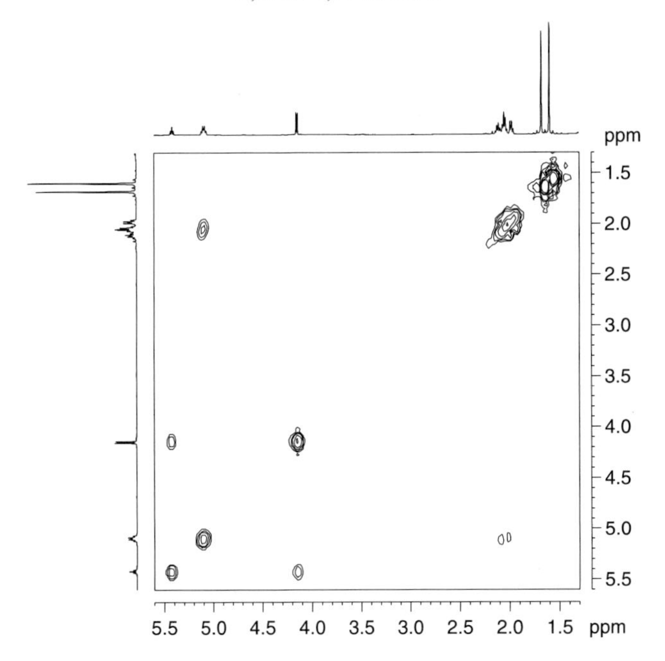

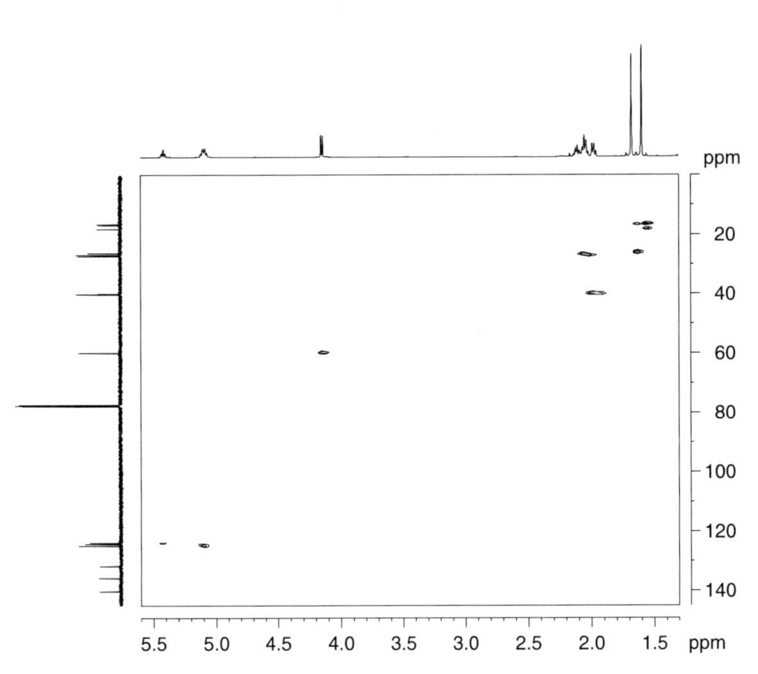

Problem 34: $C_{15}H_{26}O$
IR: 1675, 3406 (broad) cm^{-1}
500 MHz, solvent: $CDCl_3$
^1H, ^{13}C and DEPT spectra

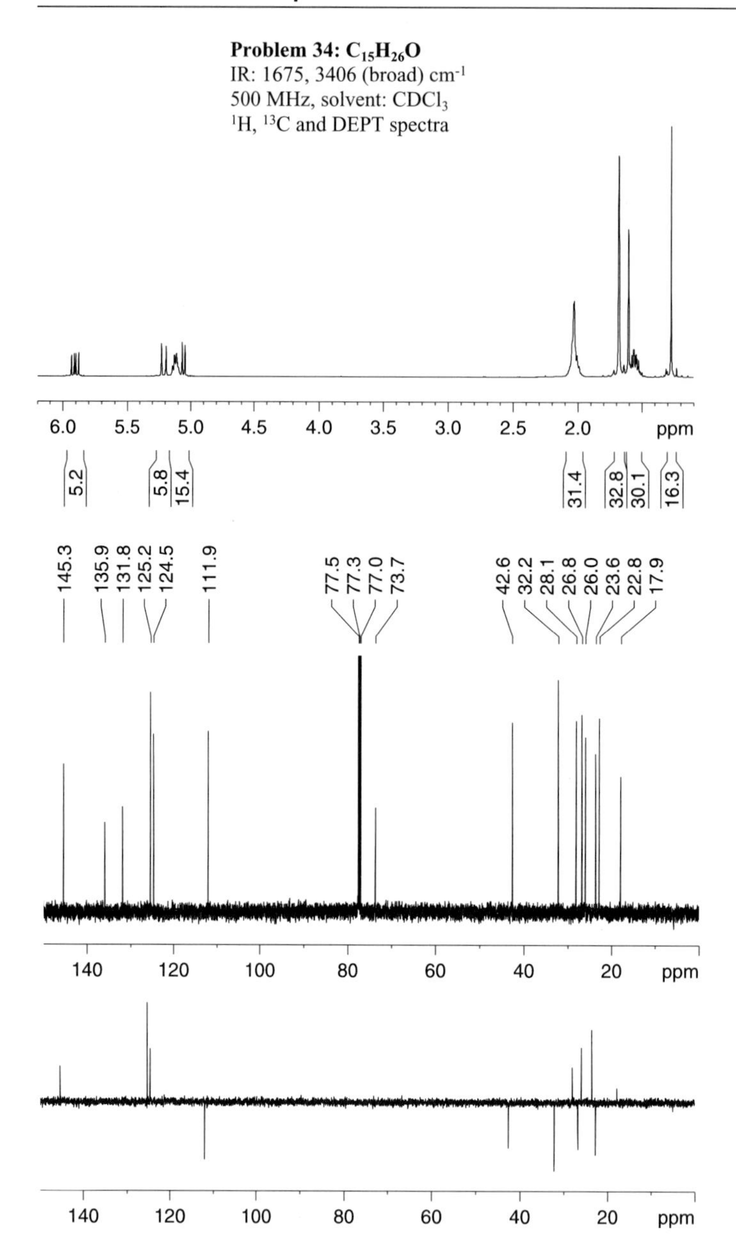

Problem 34: $C_{15}H_{26}O$
500 MHz, solvent: $CDCl_3$
H,H and C,H correlation

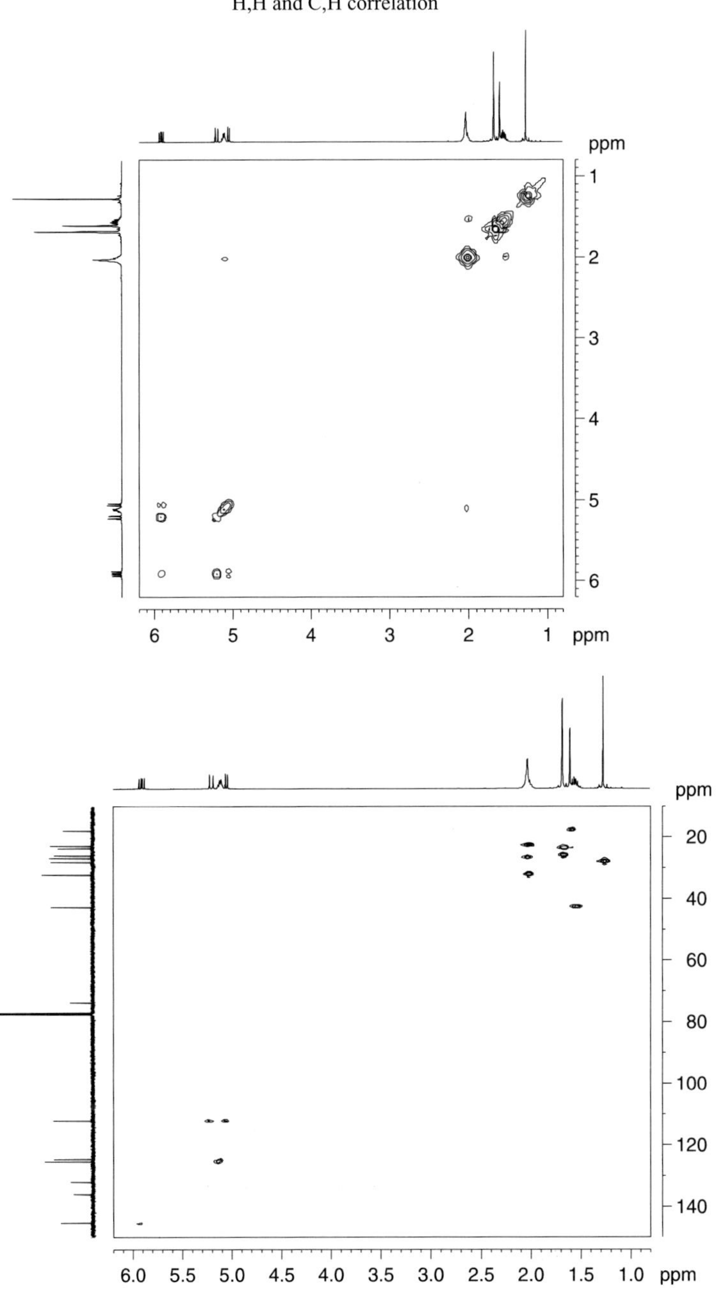

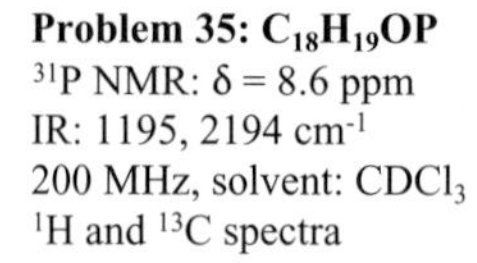

Problem 35: $C_{18}H_{19}OP$

^{31}P NMR: $\delta = 8.6$ ppm

IR: 1195, 2194 cm^{-1}

200 MHz, solvent: $CDCl_3$

^1H and ^{13}C spectra

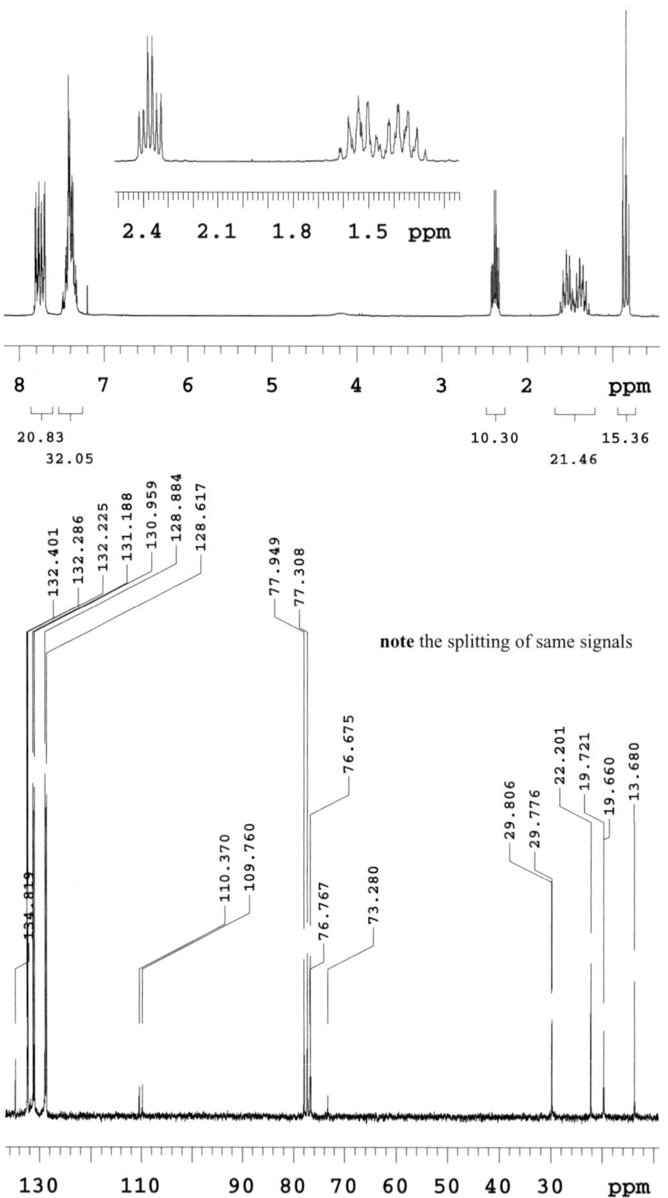

Problem 35: $C_{18}H_{19}OP$
200 MHz, solvent: $CDCl_3$
APT and DEPT spectra
expansion: peak frequency in Hz

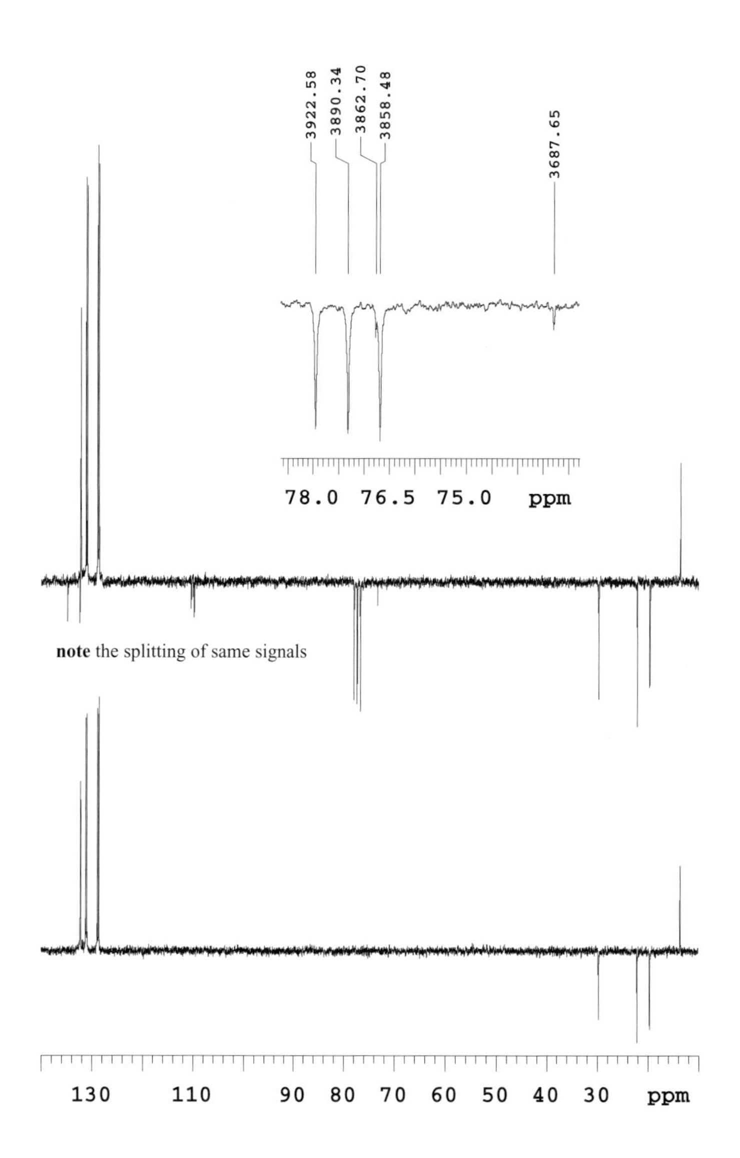

Printing: Saladruck Berlin
Binding: Stürtz AG, Würzburg